Seen Flying the Skies of NYC...
the Photos, the Theory, the Story

Published New York
(c) 2015

www.cloudofIron.com

Vol. 2

I0484863

www.cloudofIron.com   Argus Witness   witness@cloudofIron.com
New York, New York Copyright © 2015 by Argus Witness

Picture Credits: Cover: "**Nibiru Rising**" **(Jan 7, 2015,** Infrared photograph of object emerging from behind Sun through a Solar Filter "**Boy Reading on a Cloud**:" equalized , cropped, detail from "Cloud Making". "**The Gods at Work: Messages from the Dawn of Civilization**"
Photograph Mar 30, 2013 of Plane and Clouds **Processing: Edge Detection**

ISBN-10: 1511526408   ISBN-13: 978-1511626401

This book is dedicated to my patient and loving wife and children.

Thank you also to those who contributed ideas and critiques.
Also by the same author:

Imagine.... The spiritual beings that were the foundation of all the world's major religions are real and are here now, trying to warn us of impending apocalyptic times. How would they do it? Why not start sculpting some of the dangers out of clouds around the planet. Make sure our Mars spacecraft catch some pictures that turn science and physics on end. And where else to create these enormous warnings but in the modern Babylon - New York City! This monthly book contains real, recent photos taken in the NYC area, presents some ideas and ends with a fictional type of what-if story based on the real photos. Sure to be a collectors item.

The author, 30 years in the computer industry, advanced technical SCUBA diver with over 500 dives, trained as a pilot which included lots of meteorological training; is a native New Yorker. A traveler that believes that his best spirituality lies outside organized religions. He has a long history of looking up to the skies for inspiration and meaning in life.

Seen in the Skies of NYC and Mars v1.0

ANGUS WITNESS

'Enhanced' Southwestern Rock Art   Infrared Photo of Sun Jan 7, 2015

Age Unknown: Petroglyph from Saguaro National Park, AZ. Official Interpretation is of Cactus under setting Sun. But the images also look like three stars; motifs common in ancient Rock Art worldwide. Two Different Suns with a Spiral between them ?   US Parks Archives.

Infrared image of our Sun taken through a Solar Filter on January 7, 2015 5:19pm facing South from central Queens, New York City. This image has not been processed at all. Shows the Nibiru system coming out from behind the Sun. Image 1 of a set of 3.   www.cloudofIron.com

"Nibiru Rising"

There have been legends of a celestial body known as 'Planet-X' to modern astronomers, 'Nibiru' to the ancient Sumerians in Mesopotamia or the 'Destroyer' in ancient Egypt. It is a mini Solar System; our Sun's binary companion star, a Brown Dwarf Star, at its center with a number of planets. The system is on a 3,600 year elliptical orbit of our sun. It has returned to the area of our Sun and rounding the Sun, is ready to head back out again into deep space. But first, it must keep its long scheduled appointment with the Earth; placing all of us in a situation of possible peril that can be further understood by looking at mankind's historical record and the Earth's geological record.

World governments have had no bigger secret than the return of the 'Destroyer' the last thirty or so years. When it appears, it will be too late. It will look at times as if there are two Suns. The enormous dust and debris cloud accompanying it may give it the appearance of a 'Winged Disk', a symbol common to a number of ancient cultures. This is the very apocalypse spoken of in the Bible's Book of Revelation. After a brief flurry of news in the 1980's, very little was heard on this topic. An immense cover-up of information has been underway and it has worked. It has included ruining the careers of scientists that spoke on the topic. A number of astronomers have even died under unusual circumstances. No one has spoken of this except in terms related to translation of ancient Sumerian clay tablets that tell the story in detail. No precise time line for an event has been given, nor is it considered fully accepted by the 'official' scientific community that Planet-X even exists.

I started photographing the clouds, Sun, Moon and reviewing NASA photos of Mars in late 2012. In 2012 I was hospitalized with pulmonary problems for six long days. I had collapsed in a store, suddenly unable to breath. While in the hospital, I was bored and had a lot of time. I started looking at NASA's extensive websites for the Mars Missions. The files being returned from the satellite in orbit around Mars were huge, revealing very fine detail. In the first photo I examined I found many 'anomalies' including a large metallic object sticking out the side of a crater, an enormous bust from a statue, camouflaged equipment or military positions, broken down equipment, castle like ruins and other things. I was amazed.

I had also been interested in the search for 'Planet-X' and the ancient Sumerian culture for over twenty years and long ago had read the writings of Immanuel Velikovsky, a controversial author of the 1950's who proposed that the myths of ancient peoples were based on actual events. The more recent writings of Zecharia Sitchin further supported many of Velikovsky's theories and had something specific to say about civilizations such as the Sumerian's that emerged from what we now call the "Cradle of Civilization" in Mesopotamia. Now, I was hooked.

Around the same time I started simply looking up. Now, my thousands of photographs from the past few years include some that show what may be the warning signs and finally the approach of the star and planets making up the Nemesis/Planet-X system. My photos are real. Using a Solar Filter I photographed objects emerging from behind the Sun. These objects have spread out now from the Sun, with mid-March 2015 photos capturing at least one of them at the nine o'clock position relative to our Sun. I use a combination of equipment and techniques including solar filter, infrared camera, film camera, digital cameras and a lot of research.

About three years ago, something in the skies caught my attention. I was certain there were sculptures being made out of the clouds. While humans have always seen forms and faces in the clouds, this seemed different. There were entire scenes. Most seemed warnings of impending disasters.

Sikhote-alin: Soviet postage stamp depicting one of the largest observed meteor falls in history in 1947, witnessed by hundreds and commemorated on this stamp.

"The sky is falling!   The sky is falling!"
- Chicken Little, in an old children's tale.

2013 Russian"Meteor Event Chelyabinsk meteor trace

Following are some of the imagery in the clouds that attracted my attention in 2013. Note that some images because they have appeared elsewhere have references to other photos not in this book. See: **www.cloudofIron.com**

In the year 312 AD, at the Battle of Milvian Bridge, Constantine the Great saw a "Chi-Rho" style Cross in the Sky. ✳
The words 'Under this sign you shall win.' led him to have his soldiers place the cross on their shields. His army did win although smaller in size than his oppponent's. He became the the leader of Rome, allowed freedom of religion and went on to found Constantinople.
In the year 2013 AD, there appeared, while facing the NYC skyline, the Image of a huge Comet. Above and to the right of the Comet was a niche in the clouds that contained a brilliant white Cross. As I watched, the image changed as you will see in the following photographs.

Above, a comet like cloud and meteor-Skyline lightened. Below; suggested a Tsunami to me.

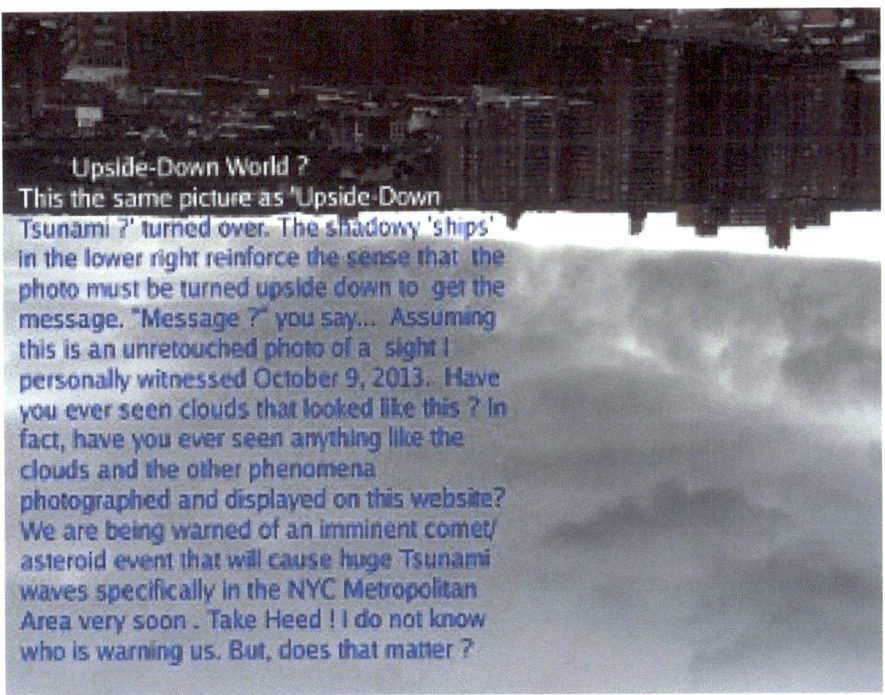

Upside-Down World ?
This the same picture as 'Upside-Down Tsunami ?' turned over. The shadowy 'ships' in the lower right reinforce the sense that the photo must be turned upside down to get the message. "Message ?" you say... Assuming this is an unretouched photo of a sight I personally witnessed October 9, 2013. Have you ever seen clouds that looked like this ? In fact, have you ever seen anything like the clouds and the other phenomena photographed and displayed on this website? We are being warned of an imminent comet/ asteroid event that will cause huge Tsunami waves specifically in the NYC Metropolitan Area very soon . Take Heed ! I do not know who is warning us. But, does that matter ?

The next photo is not processed. To me, it shows a warrior with helmet on a cloud, his abdomen pierced by a ray of light. Above him, in a boat, more men, one standing on the bow of the boat wearing a large hat.

"Preparing for Battle"

The photos below were taken in April of 2013 in the middle of the night. For hours there were changing scenes occurring in a hole in the clouds in an otherwise overcast sky. A bright light I initially thought to be the Moon was the centerpiece in these fantastic displays.

"Division of the World"

"Riding the Dark Star to Earth"

"Noah's Warning"

About 2:30am April 25, 2013 I took many pictures of the Moon. It kept changing as I watched. This is a zoomed in photo. Note to the left of his neck a word starts: 'noA' The moon's craters are washed out from the long exposure.

"The Conductor of Life"

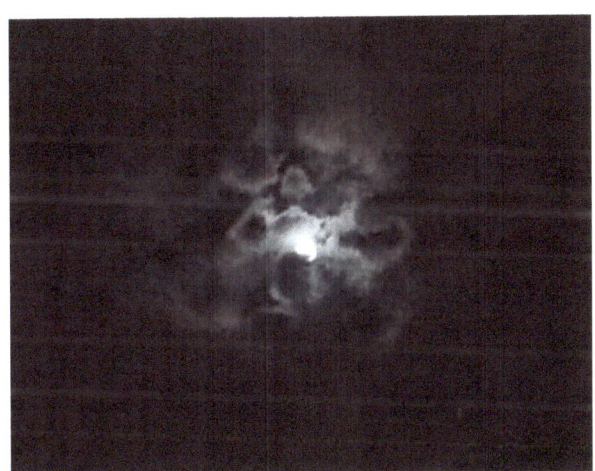

Factors contributing to a belief that Planet-X is out there:

a. The orbits of Uranus, Neptune and the trajectory of the two Voyager Spacecraft launched into deep space are all 'off' from what scientists had predicted. It was proposed that this could only be due to the gravitational effects of a large celestial body, a large Planet or Dwarf star somewhere in our Solar System, as yet, undiscovered. Scientists call this various names. If looking for the Sun's possible binary dwarf companion they call it 'Nemesis.' If referring to a large hitherto undiscovered planet it's 'Planet-X' or 'Nibiru' – the name the ancient Sumerians knew it by. For a long time, astronomers could not view the area around our Sun because the brilliance of the Sun hindered observation. For a number of years NASA has had numerous satellites covering this 'blind spot.'

Many civilizations have suddenly appeared and vanished. Science has no idea why they vanished. If humans were here millions of years and keep getting knocked back to the Stone Age, that would explain the cycles of civilizations.

    a. Egyptians that built Pyramids

    b. Mayan Culture in South America

    c. Large Structures, found underwater around the world.

    d. Generally, only large megalith stone structures have survived from the far past.

b. Odd discoveries, such as Woolly Mammoths and other prehistoric animals with food in their mouths and stomachs. Five to ten thousand years ago, they died and were frozen so suddenly that their remains were still fresh enough to have been eaten by their discoverers.

c. Ancient humans demonstrated an interest in tracking the stars as shown by megalith structures such as Stonehenge in England. It has been assumed that this was for mundane reasons such as knowing when to plant crops. Some of the knowledge required to build these megalith structures meant understanding the motions of the heavenly bodies on a timescale much longer than a human life. Great efforts were put into these observatories.

d. Many cultures around world tell legends, similar stories, of Floods and great destruction.

e. Plato, in his 'Dialogues' of Timaeus and Critias' spoke of the Earth having gone through cycles of destruction and recreation.

f. The Ancient Egyptians, Mayans, Northern Europe cultures, Hindu's, Ancient India tell legends that the sky appeared different in the past -than it does today. They also speak of cycles of destruction and Re-creation. The Hopi Indians believe they emerged from a hole in the Earth to our world, the 4th World, after the last cycle.

g. A number of belief systems; The Hopi Indians, Mayan Indians, Hindu and Christian religions all say something huge is about to happen.

h. Great underground ancient tunnel systems across Europe, Turkey and other parts of the world have been discovered. Some could hold tens of thousands of people.

i. The Christian Bible seems to predict a coming catastrophe in the Book of Revelation.

j. Charles Darwin theory of evolution has no direct evidence showing a transition- from apes to Man. Rather, the archaeological record shows sudden jumps and sudden appearances of modern man. The ancient Sumerians say that man was created by aliens; the Annunaki from one of the planets of the Nibiru system.

k. Symbols etched into rocks by very ancient man from Australia to the American Southwest show similar motifs or art. It could be that these people were recording something they saw in the skies. Symbols that permeate man's history; crosses, spiral's, even the swastika may have its basis in celestial threats.

l. In the 19th Century a US Congressman, Ignatius Donnelly, who was a writer and amateur scientist, wrote "Ragnarok: The Age of Fire and Gravel" in which he proposed the Great Flood and sinking of Atlantis were due to Cometary impacts. Very interesting reading.

m. In the 19th century, tens of thousands of clay tablets with writing on them were found in Mesopotamia. Once called Sumer, this is now the country called Iraq. This is the general area known as the 'Cradle of Civilization.'

n. None of the ideas presented herein were invented by me. There is nothing new under the Sun as they say, I am reporting what I have witnessed.

Photographs with images of things that were not seen by the naked eye. Example: "The Ancient Ones Retake Babylon", NYC Skyline Sept. 27, 2013

This has happened a number of times. The subject matter ranges from the tiny to huge things that when I see them I say to myself "How could I have not seen that?"

A couple of pages further down are night photos of the New York City skyline from the evening of September 27, 2013. I thought I had taken a series of photos of the NYC skyline on the East side of Manhattan from about 30th street up to almost 65th street. Later, when I looked at the images, there appeared to be figures, suggesting ancient Mesopotamian gods, in front of the buildings making up the NYC Skyline. There is a female, perhaps Ishtar, sitting, legs crossed, in front of the Empire State Building. She is holding a globular light in her left hand. Interestingly, the aircraft light is missing from the top of the building's antenna. I have all the original photographs. Some of the other images from that day and evening also have cloud formations and aircraft above the skyline.

Top Image: Turned 90 degrees counter clockwise and then look in the lower third of the image do you see a figure wearing a hat in the clouds?

Bottom Image: Starting at the southern end or left side of the image
30th Street – Figure Getting in/out of a large throne chair and/or hugged by a smaller female on our right.

34ᵗʰ Street – A breasted Female wearing a helmet sitting in front of the Empire State Building, her right leg over her left, holding a globe of light in her left hand.

36ᵗʰ Street – A female with long hair sitting sideways on her left thigh facing North with her right side facing us.

37ᵗʰ Street – A lantern or upstretched arm holding a platter upon which are three or four globes of light. This is almost over the head of the 36ᵗʰ street female. There is a smaller figure directly under the 'platter' holding the lights.

38ᵗʰ Street – A snake like entity coming from the West side of Manhattan, curving up and over buildings and then down again, with its head turned towards the 37ᵗʰ Street lights. Perhaps about to engage the man standing directly under the lights on a platter.

39ᵗʰ Street – A green colored monstrosity with two white eyes or observing the viewer with binoculars or glasses. It has a horizontal mouth above a large chin.

40ᵗʰ Street- As we approach the Chrysler Building at 42ⁿᵈ Street and continue north, a few vague forms including a large head facing north.

"The Ancient Ones Retake Babylon", NYC Skyline Sept. 27, 2013

On the next page, the top image is the original image, the only change being text added to it. The bottom image is an enlarged enhanced crop of the NYC skyline from about 30ᵗʰ Street to about 42ⁿᵈ Street. This is looking west from about eight miles away in central Queens. Image processing on the enlargement includes: White Balancing and Sharpening.

"The Ancient Ones Retake Babylon"
New York City Skyline 09-27-2013
Unretouched Photo (top)   7:40pm
www.cloudofIron.com

White balanced, sharpened, enlarged

(c) 2013 The Wharves Project, Inc.

The Sumerian clay tablets were not translated until early in the Twentieth Century. Many were business records, many were taken to be mythology, stories or legends. Some of these stories matched bible stories with slight variations.

A Russian scientist, Immanuel Velikovsky in the 1950's wrote a book **"Worlds in Collision"** claiming that the Earth and our Solar System had gone through periods of planetary collisions. According to Velikovsky, the evidence was there in the geology of the Earth and the current orbits of the planets. His controversial ideas were ridiculed by some scientists. Others began to reexamine their sciences. Until this point, 'Modern Science' had the opinion that changes to the Earth, such as the building of mountains etc. were the result of very gradual processes operating over millions of years. Now, some scientists began to consider a new theory called 'Catastrophism.' This theory suggested that things like Comet and Asteroid impacts were the more important events that caused the Earth to look and act the way it does.

Examples of Sumerian Clay Tablets with Cuneiform Writing on them and Sumerian Cylinder Seals:

Zecharia Sitchin and the Sumerian Tablets and Seals:
Another Russian scientist, Zecharia Sitchin spent many years reviewing the Sumerian tablets. He asked, 'What if these were not legends, but real history?' Seen from that perspective, the tablets tell a sobering tale that is very specific about celestial events. The ancient Sumerians had an uncanny knowledge of the Solar System. They recorded details that modern science did not re-discover until the 20th Century. The ancient Sumerians were quite advanced, they developed the first code of laws, and made precise astronomical observations, kept business records, and are generally credited with being the founders of modern day civilization.

Example of a Sumerian Cylinder Seal

Many of the tablets and seals show well defined astronomical subjects.
According to these 6,000 year old tablets here is a summary of one the stories they tell. (Not the 'story' of the above sample seal.)

Our Sun has a companion 'Binary Star.' (We *were* looking for 'Planet-X', it is looking even more complex than just one rogue planet.)
This companion star is a brown dwarf star with a number of planets orbiting it just as the Earth orbits the Sun.
The star is very dim and small as stars go, it is about 1/3 the size of our Sun. We'll call it the 'Dark Star.'
The Dark Star and its planets are on a orbit around our Sun that takes about 3,600 years to complete.
For half that time it is moving away from us, the other half of the time it is on a return trip to our Sun and planets.
The Dark Star; our 'Nemesis', threatens the Earth on most if not every passage. It was known as the 'Destroyer' in ancient Egypt, Lord Shiva in ancient India, and 'Wormwood' in the Biblical Book of Revelation
Some of my information, for example, referring to 'aliens' is from sources on the Internet and rightly questioned. What is less questionable, and a solid fact to me, are my photographs over the past few years of something emerging from the 'blind spot' near our Sun along with other unusual images.
One of the system's planets is huge, supposedly some four times larger than the Earth and uninhabitable. Surrounded by a huge orange/brownish debris cloud it is called 'Nibiru.'
Another of the planets, is inhabited by a civilization more advanced than ours, according to the Sumerian's recorded version of history, these extraterrestrials, the Annunaki, were instrumental in our very existence, creating the first humans. I don't want to side track my discussion of what I have photographed, but the ancient references to the Annunaki include astronomical information and background details.

You could think of the known planets orbiting our Sun all aligned, as if on the surface of a table with the Sun at the center of the table.

The orbit of the 'Dark Star' is not lined up with the other planets. It is coming up at the Sun from the floor behind the Sun.

For this reason, the Dark Star is not visible until it emerges from behind the Sun to loop around and begin its return trip out of our Solar System

It does not even have to strike us to create us great grief. The gravity of a star is huge. Just being in our Solar System is enough to cause major problems like:

1. Disrupting the Earth's and Moon's Orbit.
2. Earthquakes
3. Volcanic Eruptions
4. Weather changes
5. Tsunami (Tidal Waves)
6. Asteroid impacts (there is an enormous dust and rock cloud as part of this system)
7. Pole Shift (where the North Pole and South Poles reverse)

"The Flute Player", skies of NYC May 1, 2014. Original photo on left.
#20140501_192901
www.cloudofIron.com

A little about my interest in this topic and my lay person qualifications:

I have been interested in the topic of Planet-X for many years and coincidentally reading about the ancient Sumerians for almost the same length of time. I wrote a college research paper on NASA's search for Near Earth Objects in 2004 and got even deeper into the subject. I learned how to fly a plane and in the course of that learned much about meteorology and other aviation related topics. (I soloed 12 times, and then stopped before getting my pilot's License so I could focus on my computer business.) I previously had done freelance news photography and had a few small exhibitions and am very comfortable with a camera. I have always had great interest in history and science. Wondering many of the questions we all have; about where humankind is from and whence we are headed.

The Freemasons

I've been a Freemason since 2001 and was head of my Lodge for two terms. It was fascinating learning the little that is known about the fraternity's roots. Its most prominently displayed symbols, the Square, Compass and Plumb line are the basic tools needed to construct a building. One of Masonry's primary goals is that of 'Making Good Men Better Men', but I have always had a sense there was an even greater purpose. Without giving away any real secrets, the Mason's speak of protecting knowledge for Humanity in two separate containers. One protects against Fire, the other Flood. Why would that be? Why are our basic symbols building tools? They are used allegorically to teach lessons about character building to the Mason but there is also an implication that when civilization falls, there will be a need for knowledge to rebuild it. Officially, Masonry's biggest secret is lost. I believe that the lost secret is that of a cyclical destruction that occurs on the Earth and the Masons are an important part of the reconstruction process.

Since 2013 I have been making observations of the sky, Sun and Moon. These were prompted by seeing some incredible things in the sky that looked like intentional cloud formations and designs. Things that looked like huge tsunami waves, comets, people swimming and many other themes that seemed to be warnings. I have thousands of photos.

"Reading by the Light of the Suns" **"The Good Book"**

These images are details from various cloud photographs. Contrast and color enhancement may have been used to bring out detail bit nothing has been added to the images."Reading the Book"

During a summer thunderstorm in 2014, a number of clouds shaped like fighters lined up, and made a 90 degree turn north-east, swooping down as if to show me they were there. My building is under one of the approaches to La Guardia Airport (LGA) (Infrared Photo,ColEqualized)

Unusual Results from Using 'Neon Edge Detection' Processing "The "Wolf Cloud Marauding Queens, 2013", Original Photo

After Neon Edge Detection Filter

I sometimes apply an image processing filter called 'Edge Processing' on my pictures of clouds. Using bright neon, the 'filter' highlights details that I might not otherwise notice. If I see something interesting, I can then investigate it further. Shortly after I started using it, I saw that often the processed clouds came out markedly different than they had looked to the naked eye. Not every photos; just some. In almost every case where this phenomenon occurred the edge detection filter created an entire frightening apocalyptic scene out of the clouds. You could look at the original photo and see where the edge detection filter based its edge detection on, but there was an inexplicable additional level of detail. For example you can see, ok, that cloud became a tsunami wave and the little cloud a boat. But where did these figures (people) in the boat come from. And they are holding oars… And where did the dinosaur or dragon on the cloud come from? It has teeth and people are trying to get away from it. And on that other cloud there is something holding an open book.

For a couple of years I wondered if perhaps the programmers of the photo editing/analysis software I use had a sense of humor and had added in this 'feature' that took some simple data in the picture and used it to create these scenes.

Qualifying as a plot for a Sci-Fi movie, the other theory was too crazy to even consider – or *was* it? Maybe, the edge detection routine was actually finding and highlighting barely visible details that were being recorded by the four different cameras. Another layer of reality? Another dimension?

Here's the Science Fiction Movie theory described…
Showing up in my images were the outlines of usually invisible inter-dimensional energy fields. These were made visible to the camera by nano-sized metallic particles that were being 'Chemtrailed' into our upper atmosphere. Surely you have heard of, perhaps even seen the infamous large high flying craft of unknown origin, creating those contrail like long thin lines high in the sky. Those "Chem Trails" are rumored to be full of tiny metallic particles that are blamed by some for an attempt to wipe out humanity, for Weather Modification, Secret Electronic USAF HAARP weather warfare manipulations, providing the base material to create cloud sculptures from and my personal favorite; to simply obscure the sky and the approach of the Nibiru system. The materials first appear to be Jet engine contrails and the craft, to the naked eye, appear to be large unmarked 747 size jets. Unlike contrails whose lifespan is measured in minutes, "Chemtrails" take days to drift down. Ever expanding until the area is filled with a milky white haze through which no stars are visible. I have a few

infrared telephoto images made myself that show clearly; *those are not planes laying down the long contrail like clouds. I don't know what they are*

Imagine that!, Aliens that were invisible to us - an entire alien reality coexisting with us in real time. Until now, each civilization, silent and unseen to the other. Each, only needing those resources on Earth which they were able to perceive. A perfect co-existence.

That is, until I came along clicking away on my cameras, my recorded observations have, like the physics experiment prematurely observed by the impatient scientist, caused the impossible to become possible. Invisible realities that coexisted for eons are now aware of each other.

The tenuous thread of a connection between the two universes previously strangers to one another, it has been enlarged. Things have gone too far. Too much has been observed. Too much carefully documented.   The portal in the Time/Space continuum is now stuck in the 'Open' position.

So much for the imaginative Sci-Fi approach…
But, what am I really seeing as a result of photo analysis with these Neon Edge Processing Filters?

Recently, I tracked down one of those geniuses involved with the relevant graphic processing software's development and he assured me that it was a strictly mathematical, somewhat simple (if you know the math) piece of programming code. He showed me where I could get more information to understand it better. I am no mathematician but was able to see that the items I am seeing in the end result photos must actually be in the original scene to some extent. There are other equally scary and even more sinister possibilities:

1. Some agency or organization has updated the software on my computer to create these Images. The software is acting the same on two PC's, Mac and an Android. Have "*they*" reached *all* my resources?
2. The person I communicated with about the program was lying or is unaware that four years ago, a junior programmer inserted some 'Easter Egg' type code into the project the morning of the day he left the project. (The person I dealt with seemed sincere and was very helpful.)
3. Due to some confusion in the *spirit world,* I was given some supernatural powers at birth that were intended for a neighbor's child, so she could save the World when needed later in her life. I have unknowingly been creating the pictures psychically,   in the camera as a reflection of my own 'Dark Side." As an unfortunate side effect, in some future year when her help is desperately needed, the now adult woman 'neighbor's child' will be unable

to muster anything more powerful than the Tooth Fairy and all humanity is doomed.

So, I am left pondering what is causing the images I am getting after running the Neon Edge Detection Filter. Could they represent an alien civilization, perhaps from one of the Nibiru System's planets, attempting to warn us of tribulations to come? If so, they are only part of the picture, one of a myriad of warnings and signs that there is 'more to come,'

I have noticed some recurrent themes in the photos I have taken:
1, Person(s) reading a large book.
2. Flood, Swimming Boats,
3. Pain, Suffering, tribulation
4. Comet/Asteroid
5. Hexagonal shapes. I believe these are related to water somehow.
Here are some originals photos and their 'edge detected' end results:

"Reading the Book to a Captive Audience, 2014 (20140603_174557)"

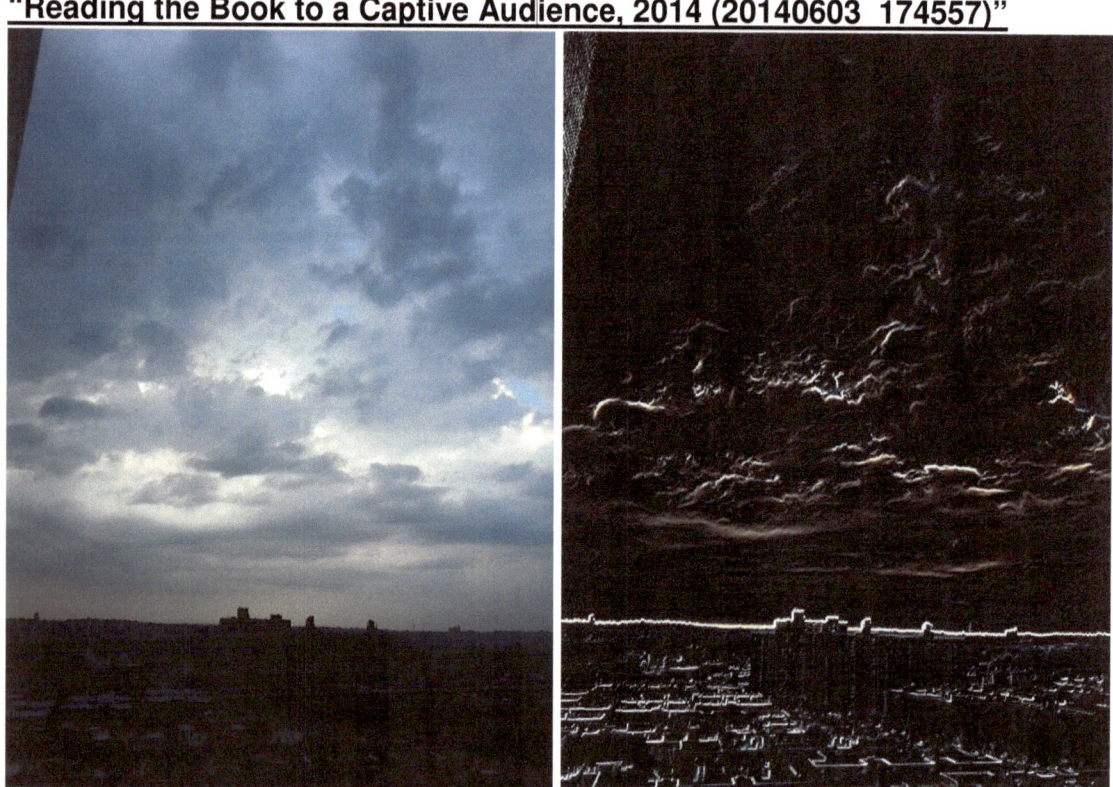

Arrival of the Crocodilian (20140427 190447 Apr 27, 2014 7:04pm

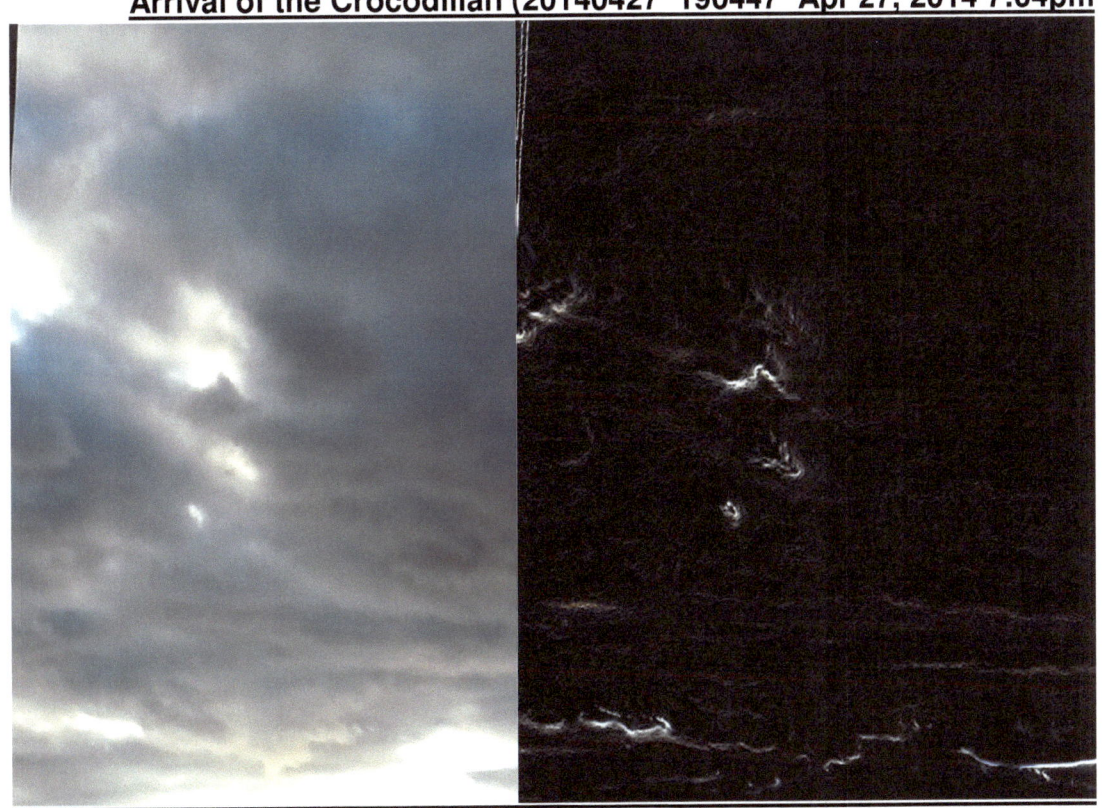

"Babylon has another visitor" (10240427 151329 5)

What is the evidence that the ancient histories of the Sumerians and others may be true ?

I had always been taking photos of the Sun and Moon but in 2014 decided to use a Solar Filter and look for changes in our satellite and star. I found them. I use the 'Burst' mode on my phone camera so it would take 20 photos in a couple of seconds. The photos do show additional objects near the Sun.. The photos appear to show that at least some of the round or spherical objects are in motion. They may be planets or stars.   There was also a brownish red smoke like haze that was all around. Ironically, my website is named 'Cloud of Iron' and that is exactly what it looks like. (see http://www.cloudofIron.com) **To be even handed with the evidence, it gets weirder; some objects look more like enormous lighting equipment of some kind rather than planets or stars. See the next two images. Until Feb. 2015, the objects appeared to be spheres. Then like lighting equipment.**

March 19, 2015 Object Photographed at nine o'clock position near Sun. See following analysis. Object is a light, mirror or huge bag of gas. Even the Sun may be a projection. Could the Earth be surrounded by an opaque envelope built up by chem trails to obscure something from us? The return of Nibiru ?
www.cloudofIron.com

Below is the same image with 'Retinex' Contrast processing…

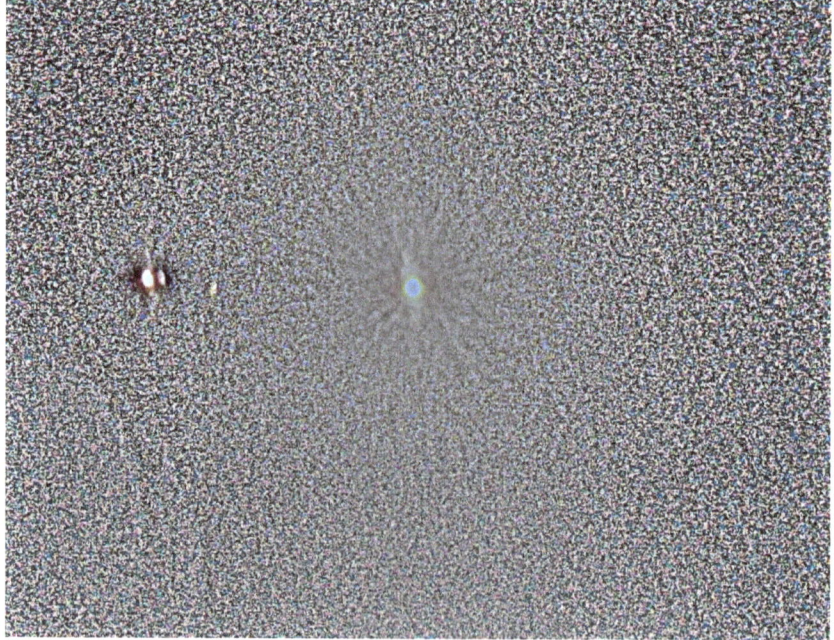

Following are some of the photos taken facing South around 5:00pm with a Solar Filter.

January 7, 2015 I believe this Infrared image may show the dwarf "Dark Star" emerging from behind our Sun.

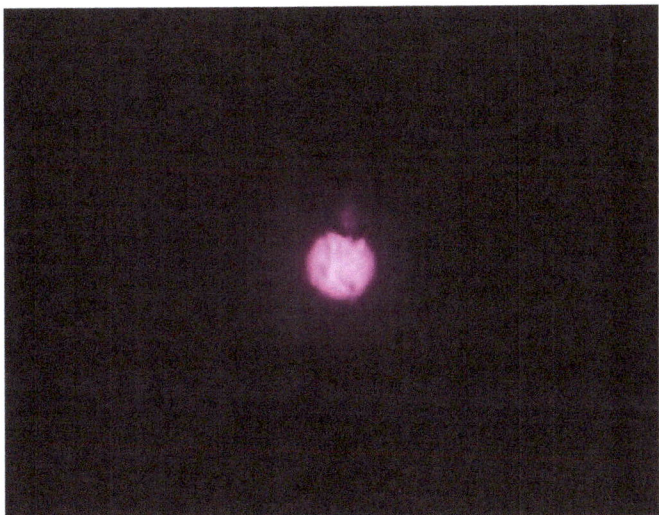

This February 13, 2015 photo below, shows the dwarf star still emerging from behind our sun and shows a large object at the 9:00 o'clock position. There are hints of additional celestial bodies at 3:00 and 6:00 o'clock. The one at 6:00 o'clock may be very large.

The planet Nibiru is said to be four times the size of the Earth.

Here are some more recent Photos, more planet/star like; all from January and early Febru
2015.

Jan 9, 2015   **Jan 17, 2015**   **Blue Filter Used Below**

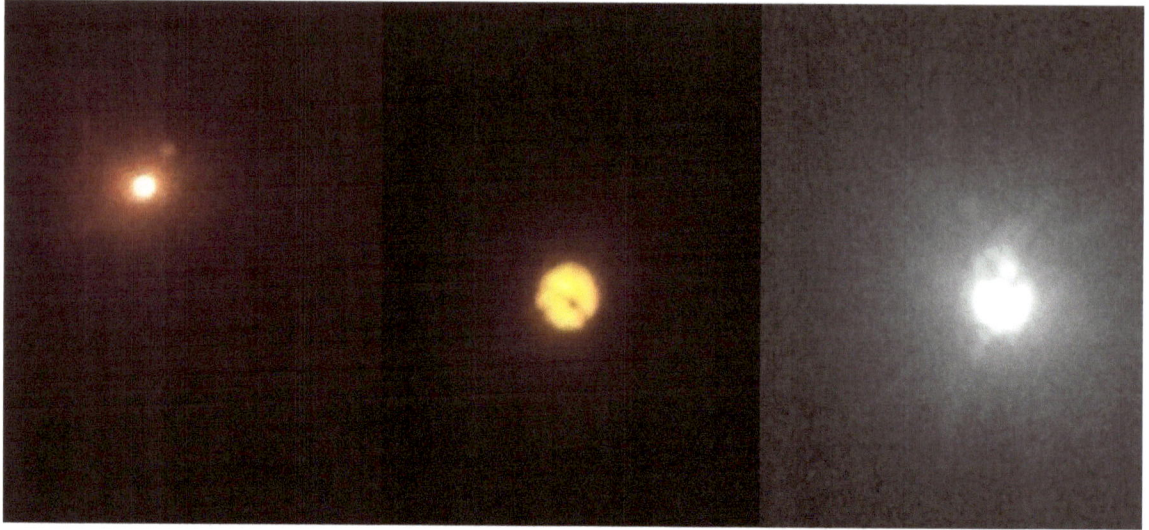

In any given three second period where I am holding down the shutter and taking about 20 photos the shape of the Sun changes greatly. It is as if something is orbiting it very fast and the object's strong gravity is creating a dark mark on the sun, bisecting it like you would cut an orange. Within fractions of a second the damage seems repaired and the Sun assumes another shape for fractions of seconds. I can only speculate, but what if one of the visiting objects were a 'black hole'?

Feb 3, 2015

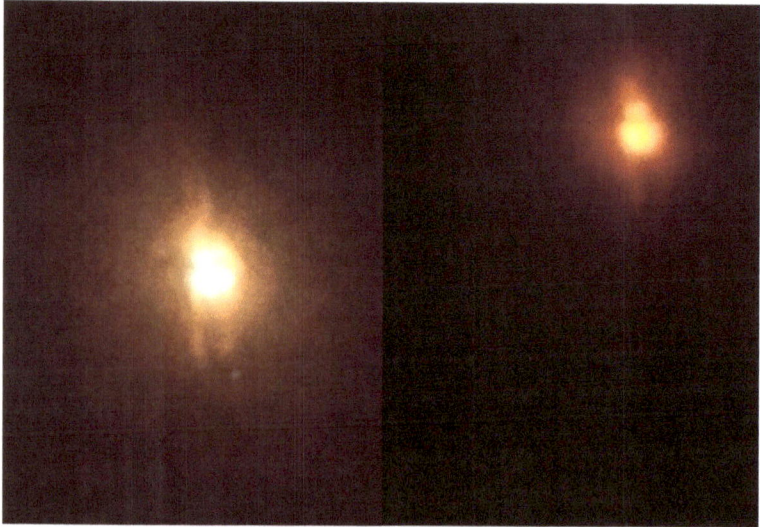

Feb 16, 2015

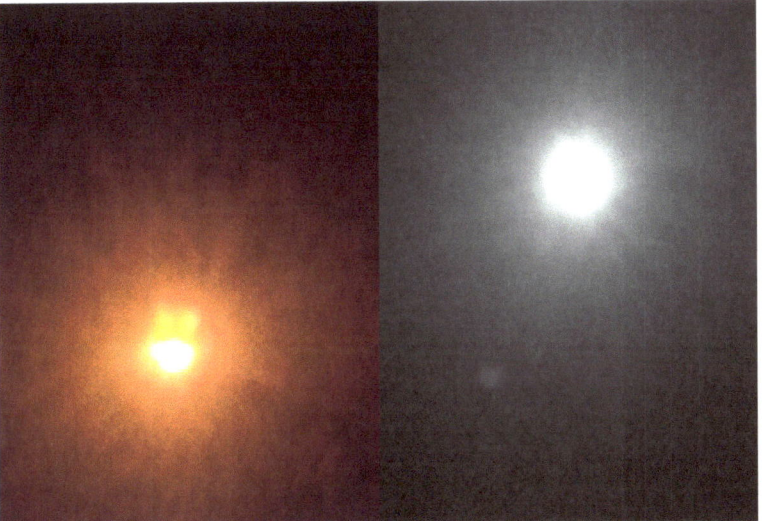

We Humans may or may not be in control of our own destiny. But we do need an awakening. Our most basic needs have been met simply enough by; food, shelter, family and society and legends of unknown provenance. Once these needs are met, 99.99% of us look no further. We live our lives as we have for untold thousands of years, quietly watching the hypnotic fire in the cave or hearth and listening to stories that we unquestioningly accept as our real history. Those in the tiny percent that try to seek more profound answers, whether in the sciences, spiritually or researching our history and human development have to really fight to get any information out that doesn't fit paradigms current at the time. They are libeled, ridiculed and many 'truth seekers' sputter out like a candle briefly lit in a storm.

Perhaps the invention of the 3-D printer recently will help jump start civilization in the coming New Age. Tribes of scribes and 'makers' and 'tinkerers' will start centers of industry. You probably need 20 or 30 machines (for spare parts), a good supply of ABS plastic and a few electric generators that run on anything that burns. Did anyone think to code the software programs to produce some basic items? What would be needed? Our modern shiny civilization, if you reduce it to its barest requirements, what would we need. I mean *really* need?

Hoes, Plows, forms for mud bricks, the arts of ore extraction and smelting, the designs of milling and metal cutting machines? Ah here's an important one: Replace all coffee machines with Melitta type paper filter coffee makers. Oh, we won't have coffee…. Let me think this out a bit.

During the last few weeks prior to the cataclysmic event, the government may try to retain some control over people that have no control by relocating them to 'FEMA Family Centers and Camps.' It is from here that will come, the FEMA Survivors that will be the source of labor desperately needed to build the Fifth World of the Hopi Indians. Will the survivor children born in the FEMA Camps work as equals alongside or be slaves to the emerging tunnel/bunker survivor population? I think the latter.

Will they in turn find that the surface has been claimed here and there by groups and tribes of Survivalists and that every inch of land must be fought for ? Or, perhaps an alien race has been coveting our planet and waiting for the coming of Nibiru to make their move. Perhaps all human groups will emerge into a world where they are equally slaves of a civilization that pays as much attention to their needs as humans did to smart domesticated animals. It may be true that a New World Order had been in the planning since the early 20th Century; did anyone guess it would be like this?

Great efforts have been made to keep the impending arrival of our Sun's unwelcome relative a secret. To keep it a secret until it's too late for anyone to do anything to save themselves has been a herculean task. The technologies required are mind boggling and call into question whether it's within human ability to perform some of the sleight of hand necessary to keep the entire population of a planet in the dark. Is an alien civilization and their agenda, guiding our and assisting our government in managing this event?

Many of the warnings have been done by sculpting the clouds. I don't know if humans can do this on their own or had help but I saw my first 'UFO' in 2013 and have seen many since. You may have heard of Chemtrails. One theory is that this persistent contrail like spraying in the upper atmosphere of nano-particle sized metals such as Barium and Aluminum provides 'fake' cloud material with which drones that are using advanced cloaking and propulsion systems can sculpt the clouds. From the ground, it appears that these 'Chemtrails' are being sprayed very high up by large familiar human aircraft.

May 1, 2014   Chemtrails being turned into Sculptures

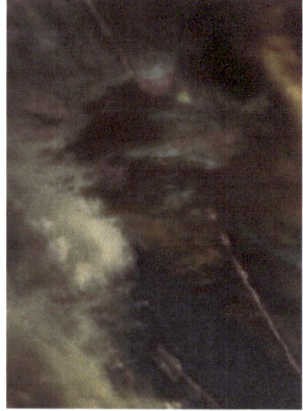

"Angelic Mother and Child" Sept 7, 2014

"An Angel leaving her troubles behind her."

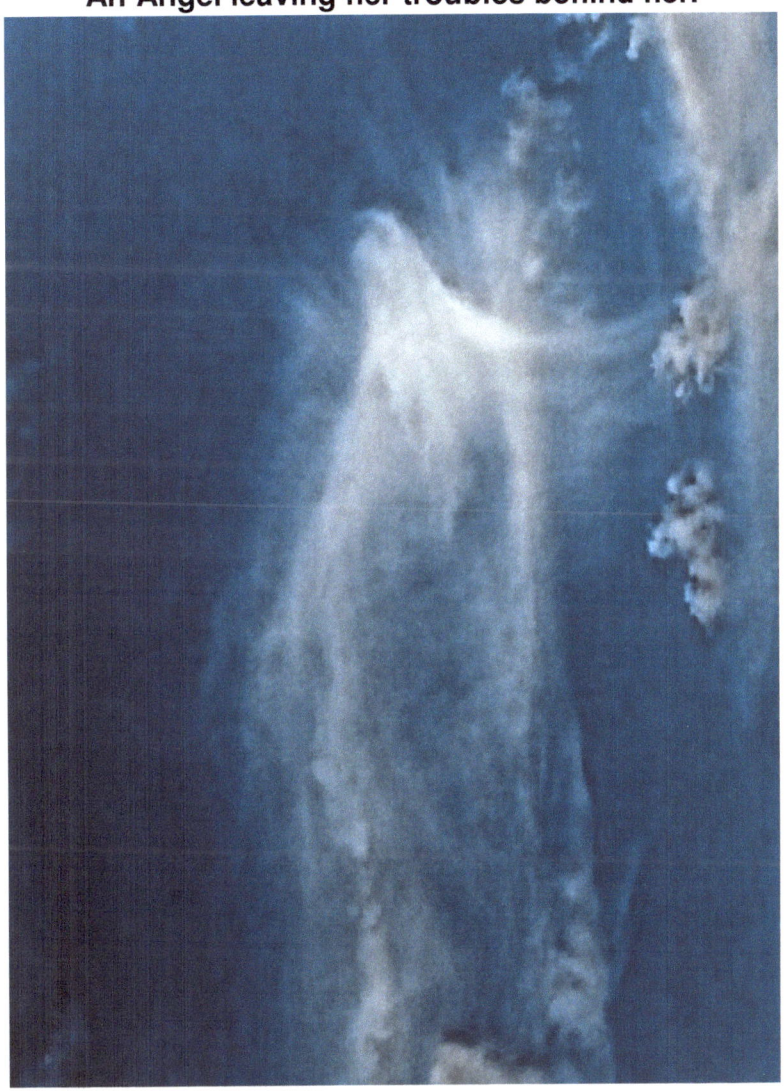

Cloud Making and Holographic Projection: March 22, 2013

| # 1 | # 2 | #3 |

'Face in the Sky' July 17, 2014   Skies over New York City.

Is this an image of a person, possibly a familiar one, gesturing with their left hand as they speak? In the Background? In the lower right of photo #1 is a possible image of an assailant.

On the next page is a photograph taken March 22, 2013. It has a number of interesting features:

1. Upper left. UFO (dark box shape) generating cloud material.
2. Center, something, assumed to be another UFO, hidden in the clouds billowing out cloud material at a very high volume.
3. At the three o'clock position, a color, 3-D image of someone sitting on the cloud reading a book. This was not seen by the naked eye at the time of the photograph.

A UFO is an Unidentified Flying Object. No one said they are aliens. It could represent a secret human technology.

The second image is the exact same photo after processing with an 'Edge Detection Filter.'

CloudMaking1__20130322_171851.JPG
Unprocessed Version:

CloudMaking1__20130322_171851.JPG
Processed with Edge Detection Filter (GIMP 2.8 9/.3)

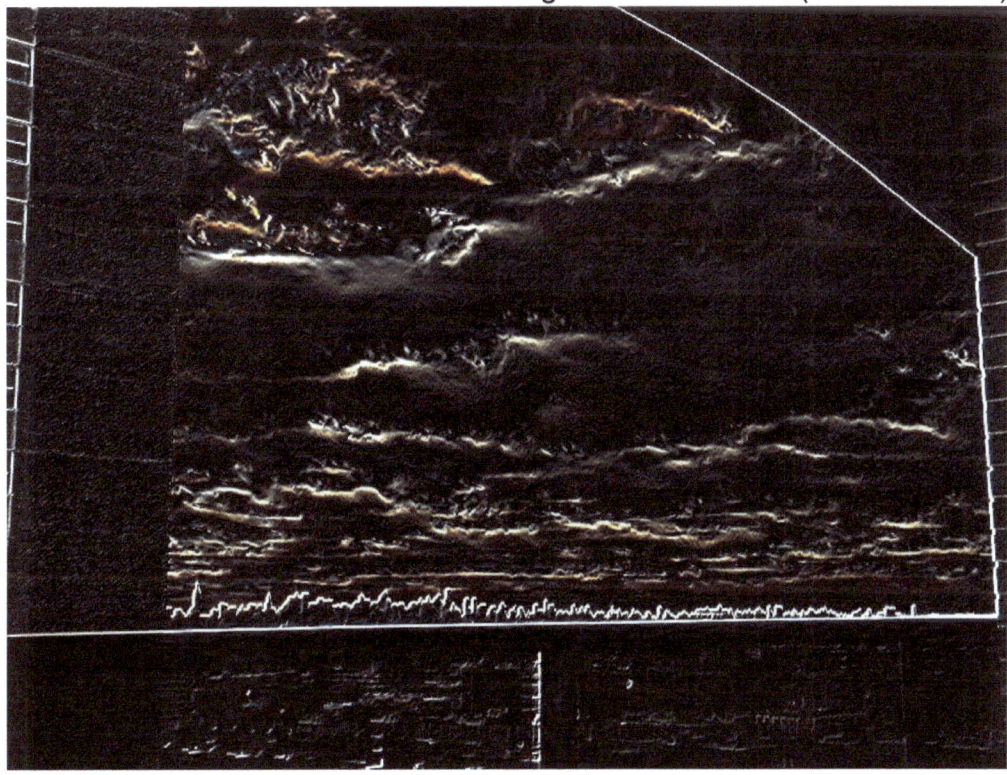

Magician's Moon

I have twice seen imagery suggesting a Magician manipulating celestial bodies. In reviewing NASA Mars photos returned from the Curiosity Rover mission, I inverted the color of the site of the first drilling into Mars and the picture surprised me. It was a Magician wearing a dark cloak and a red 'Top hat' holding what I thought was a crystal ball with what appeared to be a human form inside being pelted with rocks. I took this to be related to a future Comet or Asteroid strike. I now see that the bright yellow circle (the drilling hole) could maybe represent the Magician making the Sun disappear.

The second Magician was also an inverted photograph of the Sun, there were some clouds around it. When I looked at the photo, I was surprised as I often am. It was an image of a Magician wearing a black top hat waving his hands over the Sun or Moon; as if saying "1,2,3 Gone!"
One of the effects a visiting star would have on us is to disrupt our orbit. Previous histories of the approach of the Dark Star specifically mention things like the Sun rising in the West and setting in the East a couple of times.

As mentioned earlier, In April 2013 I watched a show in the nighttime sky where a very bright light played a central role. That light evidently was either supposed to represent the Moon symbolically or was used to make the Moon 'do tricks' while the real Moon was presumably behind heavy cloud cover over New York City. At the start, I thought that the bright light was in fact the Moon.
After watching a while I thought I had it figured out and that the real Moon was behind the clouds and this was an incredibly bright lamp of some kind.

In August of 2013, a similar show was presented by agents unknown that featured the 'Beast of Brooklyn' as I called the snake/dragon like head in the sky with open mouth that used the 'Moon' as eyes.

       'A'   **B'**   **'C'**

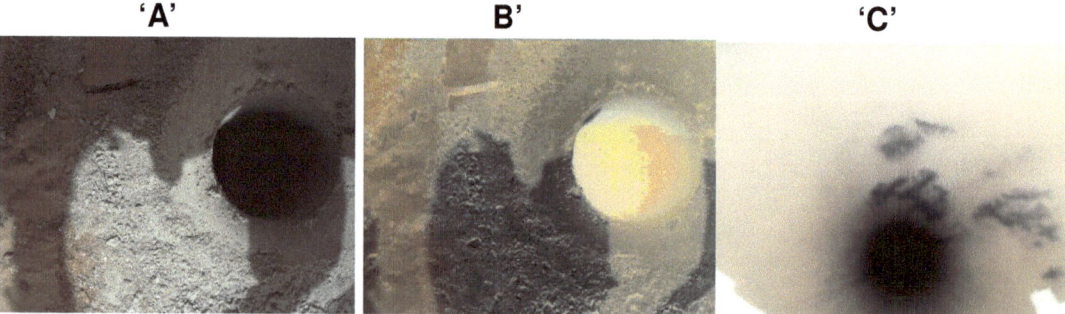

Mars: 'A', is NASA's photo #0182MH0150002011E1_DXXXtaken by the Curiosity Rover on Feb 19, 2013 of the Rover's drill hole in the Martian Surface. 'B' is photo 'A' after I applied 'value Inversion' to it. 'B' seems to show a Magician with an orange top hat on, wearing a cloak, holding the Sun as if a crystal ball. 'C' is an inverted photo of the Sun July 18, 2013 showing a Magician with a top hat making the Sun vanish in an act of 'prestidigitation.'

Beast over Brooklyn

Early morning hours of August 18, 2013 this beast could be seen in the southern skies over Brooklyn, New York. Just like back in April, The scene was constantly evolving. I stood there on the balcony marveling at the technology that created such displays. I also wondered how much more time the modern Babylon has..... This is only Color enhanced photo in set.

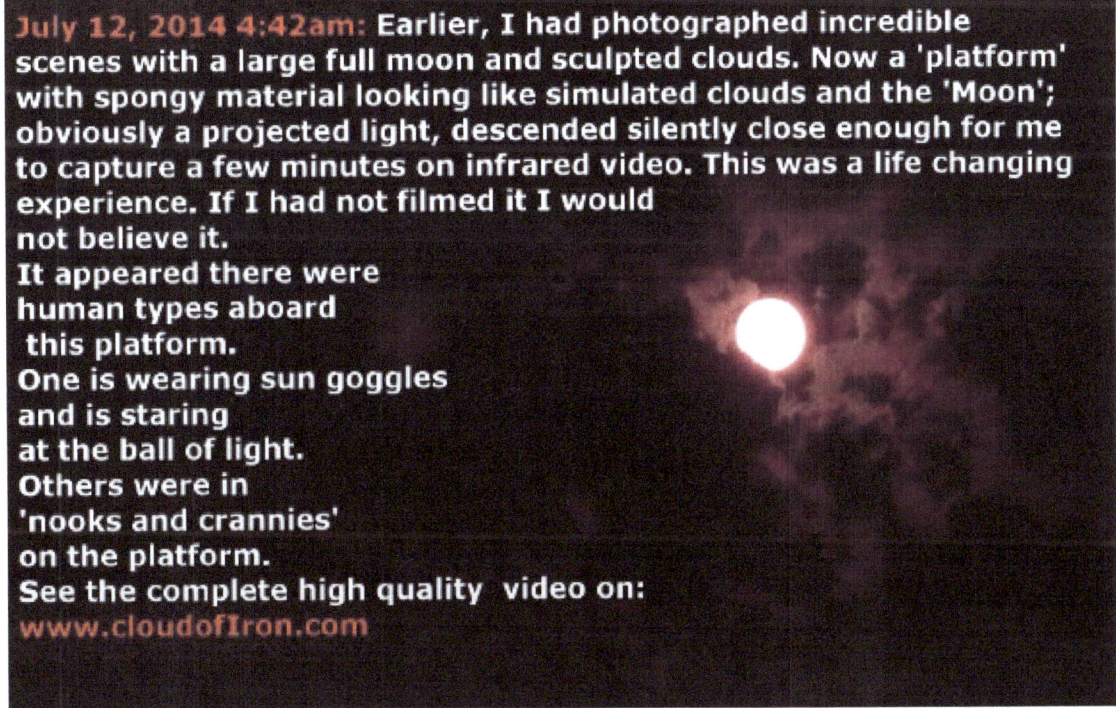

July 12, 2014 4:42am: Earlier, I had photographed incredible scenes with a large full moon and sculpted clouds. Now a 'platform' with spongy material looking like simulated clouds and the 'Moon'; obviously a projected light, descended silently close enough for me to capture a few minutes on infrared video. This was a life changing experience. If I had not filmed it I would not believe it.
It appeared there were human types aboard this platform.
One is wearing sun goggles and is staring at the ball of light.
Others were in 'nooks and crannies' on the platform.
See the complete high quality video on:
www.cloudofIron.com

Meanwhile, in general, I started looking at the Moon a bit differently. Sometimes it appeared as a slightly wobbling image with edges less defined than normal and details were not clear. Sometimes there was a 'wave' of color across it, similar to the view on a hot summers day as the rising heat from the road makes the image a bit wobbly. I attributed this to aging eyesight.

On the evening of Jan 29, 2015 at about 1:30 am, I looked out and saw to my surprise two partial moons in the sky. One was very bright with no details. Immediately behind it and slightly above it was a second, the exact same shape but it was the Moon – craters and all. I snapped a few pictures, the images in the sky eventually aligned and the moon crossed the sky at what appeared to be its normal speed ending with a beautiful huge orange Moon that came gently to rest in the West.

I thought about it a bit and decided that I had seen (and photographed) a holographic projection of the Moon. From my limited understanding of 3-D holographic projections, there are two projections that have to be aligned perfectly for the effect to work. One provides the image's brightness and the other the detail. I had seen a briefly misaligned projection. Now, why would anyone want to project an image of the Moon? Don't we have one? Or perhaps, we don't anymore, or, it's in the wrong spot due to the Dark Star's gravity?

The two images below are unaltered from the night I took them a. Photographs were taken with both a regular camera and an infrared camera.

Double Moon over NYC   1/29/2015   **Infrared Double Moon (same night)**

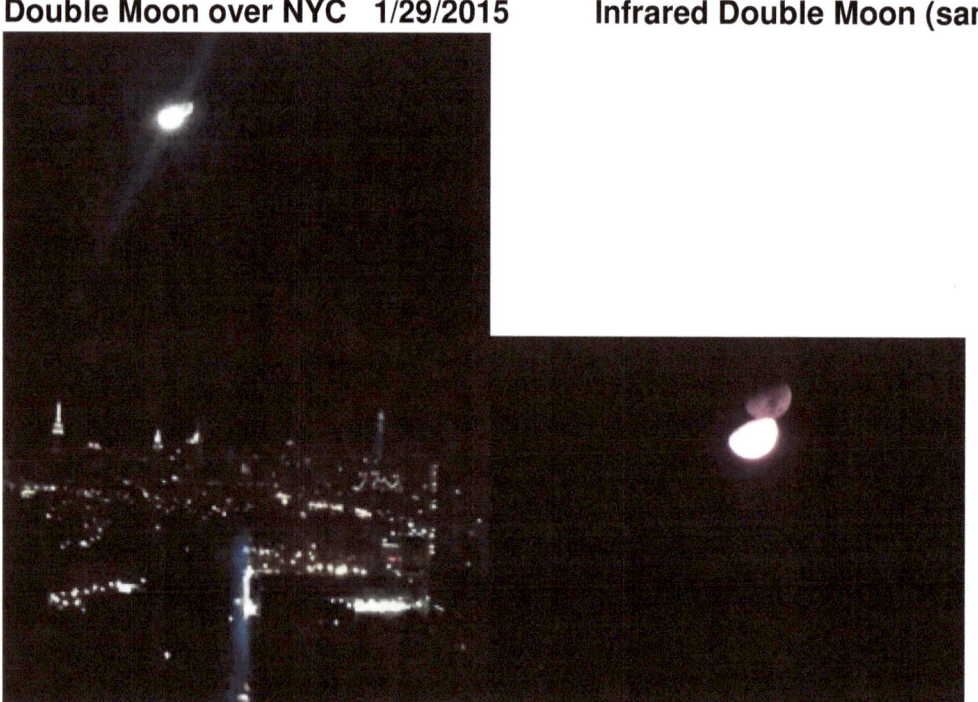

I decided to run Neon Edge Detection on the Photo of this double Moon and got the below result. Hmmm, that seemed familiar. I pulled out old photos of the Sun I had taken in 2013. I had done some edge processing of our star back then.

SUN
June 28, 2013   **Edge Detected Neon**

Are there similarities in the strange Sun and the Strange Moon. ?

MOON
July 12, 2014   **Edge Detected Neon**

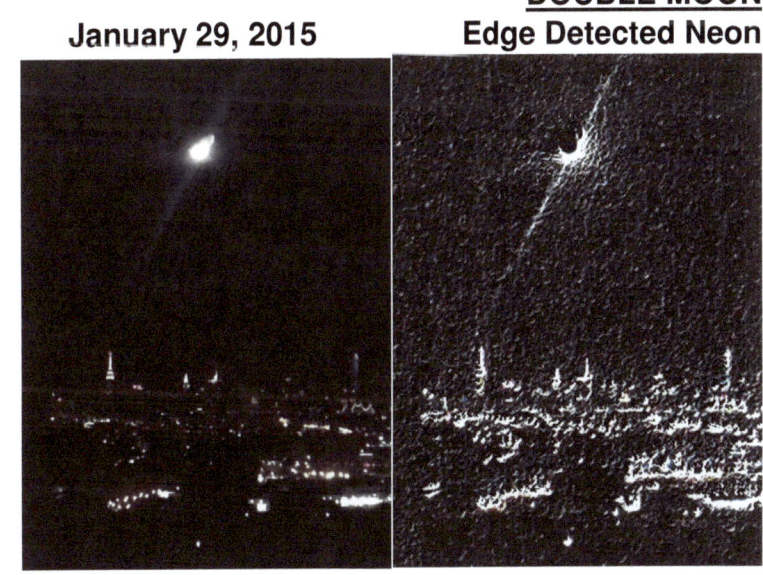

DOUBLE MOON
January 29, 2015   **Edge Detected Neon**

There are some similarities between the Edge Detection processed images of the Sun and the Moon. The biggest similarity is that for both the Moon and Sun, the edge detected version has a break in the circular shape from about 9 o'clock to 12'o'clock. The 9 to 12 break in the image may very well be an artifact of any bright spot after edge detection analysis. Are both heavenly bodies projections and the similarity is due to an artifact of the equipment producing the effect? My experience and photos of the double image Moon indicates to me that in spite of my disbelief and common sense notions that the Moon can't be projected; that our Moon was a projection that night and it does whatever it feels like! Who is doing this and why? Can we pick what we want projected? OK, my turn…. I think I want Nibiru and its *system* to be a projection. Is this projection stuff even possible using current Human technologies? Why would it be done? I can think of only one reason. Here again we are off in the realm of Science Fiction movie plots. The Earth is off its normal orbit and someone doesn't want this to become common knowledge. Why would that be? Has the Nibiru system already pulled the Earth out of orbit and between now and when the real action starts, the less people that know – the better? Maybe we can check for the location of celestial objects in Google Sky and similar programs. Maybe that won't work.

Earth's Moon is a quarter of a million miles away. The majority of Earth's cloud activity occurs within a few miles of the planet's surface. Cloud's behind the Moon ? Then it's not the Moon. July 12, 2014 3:48am Central Queens New York.   www.cloudofIron.com

"The Moon: A Problem is Discovered"

March 23, 2015 at about 10:15pm a problem was discovered with the Moon. The appropriate notifications were made. Protocol dictated immediate cessation of the projection. While awaiting authorization to make a risky live adjustment, 'Feet on the Ground' , or rather 'In the Air'   decided to proceed with the fix. The adjustment did not achieve the desired results, and the projection of the Moon was turned off and the craft left the area. There was a brief period as the projector headed uncloaked in a Northerly direction that the image of the Moon was still visible. Where is the real moon ?

Image to the left shows the issue with the Moon a few minutes prior to photo on right.

www.cloudofIron.com

That is the original photo above with an enlarged Moon and 'Moon Tender' aircraft to the right. The 'Tender'; a soundless, no visible means of propulsion flying craft consisting of a metallic looking cylinder and fork like appendage holding a light source. The light colors, are very similar to what I've seen in the Moon and like it or not – the Sun too. There are also similarities to other UFOs.

To hide that there is a big bad thing about to hit the Earth or that the Earth's position in space is not where we normally are, Google Sky and other telescope apps would have to have their programs and/or data be altered so as to not give away the fact. That's possible. I have an early model GPS star finder. After initial setup fifteen or so years ago, I never updated it on the computer. Maybe if I tell it to point to a few planets and stars I will be able to compare that result with the result from Google Sky and find out if I'm right. A preliminary test where I asked both devices to point to Jupiter resulted in a difference of almost 90 degrees in where they said the giant planet will be. I need to test more with a few different stars and planets.

Docile, Apathetic, Unquestioning, Sheep

There are many conspiracy theories out there. A number of them have to do with mind control and chemicals being used to turn us all into docile, apathetic, unquestioning 'sheep.'

The chemicals include Fluoride placed in drinking water; ostensibly to strengthen teeth. The metals such as Aluminum, Barium, and other Nano particle sized ingredients in 'Chemtrails' are linked to brain disorders such as Alzheimer's and are said to be part of a plan to control humanity. Nano particle are so tiny, they can easily traverse the 'blood-brain barrier.' Much research is done on how Nano particles can deliver medicines across this historically difficult brain defense. Allegedly there is an increase in Alzheimer's disease linked to the persistent trails in the sky. We'll revisit 'Chemtrails' shortly.

But it is the apathy, medieval styled closed mindedness and unwillingness to even look around that has me the most concerned...

How is it possible that I am the only one that saw and photographed two moons in the sky with two cameras?

How is it possible that no one has noticed problems with the Sun?

Nobody else has used a telescopic lens on a Chemtrail plane?

No one sees unusual cloud formations?
There are a few people reporting these types of sightings on YouuTube but I would expect more. Most people still look at you funny if you try to talk about aliens, UFOs etc. "Do you believe in Aliens? ", I am asked as if it is a matter to have an opinion about; like "Do you believe in Santa Claus?" and the Tooth Fairy. Well, I know more than one very

intelligent person that is certain that in our infinite universe the existence of other sentient life forms is doubtful, but they very sure and have faith that someone was brought back from the dead, walked on water and did other 'impossible' things. Seems to me like aliens and cloud designs should also be possible.

Sometimes I get an indifferent reaction when I show some impossibility I captured on the camera. When pressed, the person admits not caring "So?, Here, look at a picture of my car."
I'm still asking the question" Why are we here? "; and I have no more of an answer now than I did 40 years ago.
Are we living our lives somehow 'in-transit' on our way to a higher existence? Are we the results of experiments by Extra-Terrestrials looking for the perfect slave race? Are we actually working for them, benefiting them in ways we don't even understand? Were we created by 'God' and placed here to grow spiritually? Wouldn't God be *the* most advanced spiritual being, but still just the highest form of extra-terrestrial?

I still have *all* these questions but the big one for now is "Why is the general population not taking note that something *Big!* Is going on now? Whether the signs are by God, Man or Alien does not matter. Something very big is about to happen.

Right now, in New York City skies are objects similar to the one in the photograph at the right. They have taken up station at fixed positions over the city. They silently maintain their position, a flashing, and ever playing display of lights emanating from them, there is a decidedly religious look to this one, a TEKEL-color Moses having received his mandate from God, coming down Mount Sinai with the two tablets in front of him. He pauses halfway down the mountain gazing at the people. Have they been weighed and found wanting? Will this silent sentinel dash the Ten Commandments to the ground? These machines hover effortlessly like a helicopter, easily changing direction yet they land like planes. They have an

4-28-15 UFO IR NYC

uncanny ability to just vanish, I imagine by activating some cloaking device that renders them invisible. All these technologies suggest off world origins. Yet, they have red and green aviation lights suggesting Earthly origins. Actually nobody, whether human or alien wants a mid-air collision. Are these manned with humans or aliens? Are they drones with a number of them operated from a manned control unit also hidden in the skies? I see them on the clear nights, either hovering or flitting about on some important task. Some nights, they get bored and put on a show. One recent early evening, three of them danced together on a cloud flashing their colors in a synchronous song of light, against the purple magenta skies of a sunset. The nights I can't see them, New York

City is enveloped in a pea soup fog of nano particulate matter from Chemtrails. Whatever is in Chemtrails, we are all absorbing it. The tiny tiny tiny materials in Chemtrails have made their ways through our lungs, through our food, through our skin. Is it possible that there are so many reflective particles in the air that they could cause a regular light to take on a twinkling effect?

The Chemtrails Sprayer Craft are very fast. To the naked eye, they look like big lumbering jets with two engines. The speed with which they cross the entire sky is another story. They move across the sky so fast, yet are very high up. How fast does a plane have to be to cross the entire sky in 30 seconds? We can easily see the Chemtrails crisscrossing our skies. Like the "Tholian Web" episode of the original Star Trek television series, our world is being made a bit smaller each day, each hour as the web around our planet is completed. What if there are extra-terrestrials *and* there is something hurtling towards Earth, returning every 3,600 years. What if, we meet up with this alien race periodically when their planet returns to our part of the Solar System? We have known them as 'God.' They have compassion for us and use 'Chemtrails' to weave a new *'firmament'* around the Earth every time this happens; like the shell of an egg, so by not seeing, our suffering is limited to a few minutes or days (unless we survive).

Rebuilding the 'Firmament' of the Bible

October 8, 2014 6:29pm Chemtrail Sprayer over New York City. To the unaided eye this appeared to be a high altitude large jet aircraft. Zoomed infrared image. www.cloudofIron.com

Chemtrails - These are apparently from the jet exhausts of large military and commercial planes. Observed worldwide, the trails are said to contain nano-size metallic chemicals that are supposed to be part of a global program to reduce global warming. Taking days to drift down, they are used to create cloud sculptures and more importantly, to create an obscuring envelope around our planet to hide something.
Could it be the legendary   Nibiru/Planet X making its 3,600 year   return trip to our part of the   Solar System ?

Zoomed photos of the aircraft involved reveal odd looking aircraft. In fact, they only look like aircraft to the naked eye. At least some of these are not planes. Is some other more advanced civilization seeding our skies with something? Is human technology this advanced ? Are they working with human governments? Is this for the benefit of the average human on Earth ?
www.cloudofIron.com

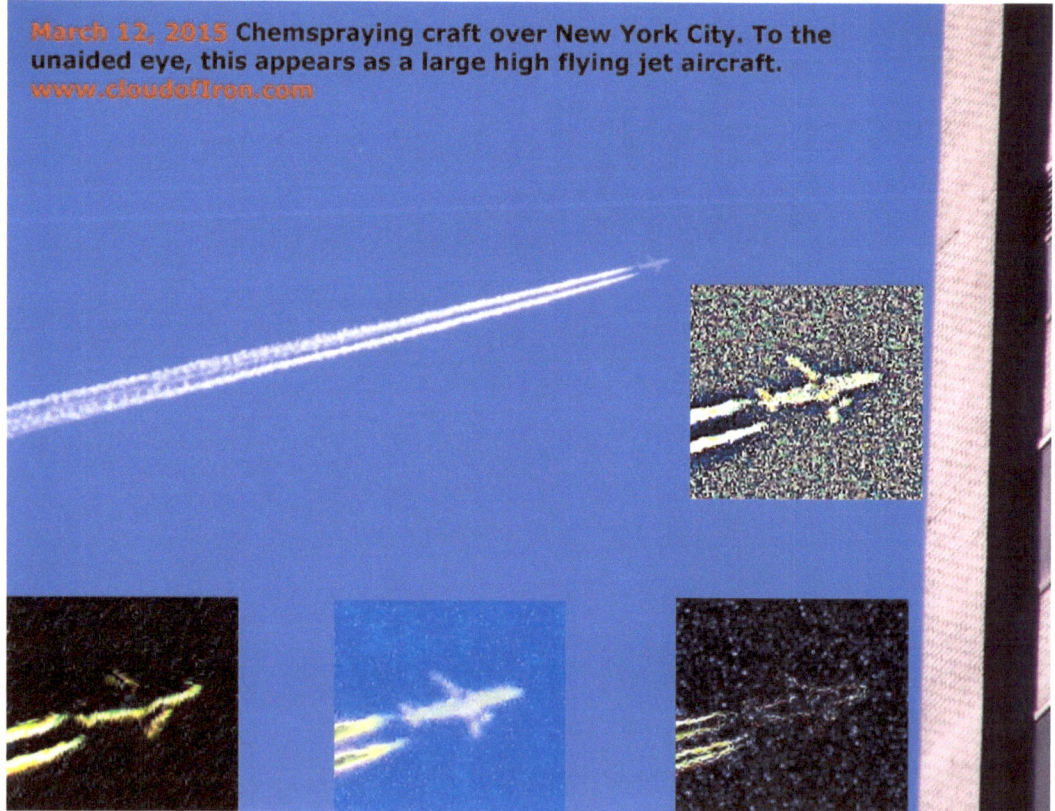

March 12, 2015 Chemspraying craft over New York City. To the unaided eye, this appears as a large high flying jet aircraft.
www.cloudofIron.com

June 5, 2014 9:14pm Chemsprayer Craft Over New York City. Craft had stopped spraying when infrared photo was taken. To the naked eye, appeared to be a high flying large jet. See zoomed inset below. www.cloudofIron.com

October 8, 2014 6:34pm Skies over New York City. A second Chemtrail Sprayer, this one better concealed as a large jet aircraft high in the sky. Close examination reveals, like the others, it's not right. One of the wings, in this case the left, is either not really there, or not 90 degrees relative to the rudder... www.cloudofIron.com

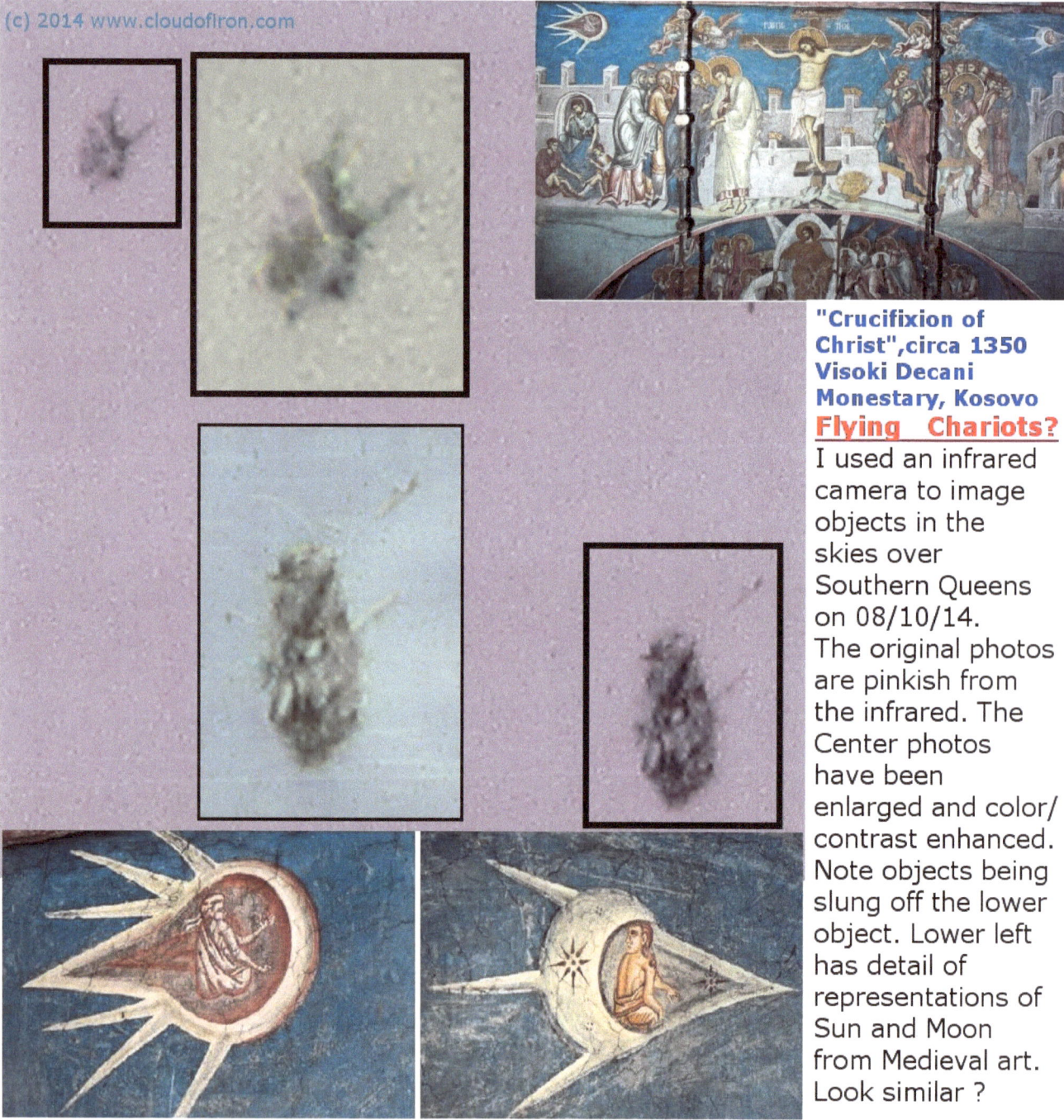

"Crucifixion of Christ",circa 1350 Visoki Decani Monestary, Kosovo

Flying Chariots?

I used an infrared camera to image objects in the skies over Southern Queens on 08/10/14. The original photos are pinkish from the infrared. The Center photos have been enlarged and color/ contrast enhanced. Note objects being slung off the lower object. Lower left has detail of representations of Sun and Moon from Medieval art. Look similar ?

Here are some more Edge Detection Images

2013-Mar-30 Dinosaur and Boats
www.cloudofIron.com

2013-Aug-05
"behind the curtain"
cloudofIron.com

2013-03-30 In the skies above New York City, following the approach into La Guardia known as the 'Expressway Approach", a large plane like object is given a checking out in Frame #2 below. This is revealed by edge detection image processing. The original photos are on the bottom. Is it a plane ? Who is checking it out ?

#3 Bottom Photo, "Plane" on left, An arrow on right. The arrow points not to the plane but to a ball shaped cloud between them. Location of an asteroid event ?   www.cloudofIron.com

Below is a 'edge detection' processed photo (March 30, 2013 1:50pm) of a commercial plane on approach to La Guardia Airport. It is in front of a cloud. Would you have known that ? This looks like a painting by Nicolas Poussin of a plane with downward and rear pointing powerplants that might be jet engines. It is being grasped by an enormous being on the fuselage right before the wings. A funnel shaped fitting on the end of a long pipe is transferring some substance (It may be blue; H_2O ? Plasmoid state energy?) The being is wearing a hat and wearing something on his left wrist with a yellow dial. The ancient Sumerians depicted their gods in a similar manner on cylinder seals and in their art. The pipe leads upwards becoming part of and vanishing into a large winged being. Everything in this picture of some clouds and a plane seems to be designed to speak to our most primeval instincts, awe and fears. These are images of the gods at work - almost exactly as described at the dawn of civilization.

Other design elements in the original larger photo include a spear (arrow) shaped cloud pointing at a spherical object. You'll see that possible warning of an asteroid later in this presentation. I believe we are looking at members of a highly technologically advanced civilization that has mastered the details of a quantum universe we are just now begining to discover. Initially I thought that these types of images were 'artifacts' or 'creative programming' of Edge Detection programming routines. Based on looking at hundreds of ad hoc spontaneous photos I have taken over the past few years, I think that these 'artifacts' may be real. We only see in a limited range of wavelengths. www.cloudofIron.com

If you take off Night Vision equipment, everything you just saw is still in existence - you just can't see it. There is no question that objects viewed on infrared equipment are 'real.' Of course they are there all the time - you just can't see them. They are very real and very solid. Anyone who has a pet has seen them bark at something that is not there, or appear to be stalking or act like someone is at the front door. You, might have even checked the door and found no one there. Many animals have a broader range of visible light they can perceive than humans can. I think that science will discover that life is all around us, whether invisible to our 3-D senses or, coexisting in a multidimensional space normally each unnoticed by the other .

What then, is the Edge Detection programming picking up in some photos ? I corresponded with someone who knew about the inner workings of the algorithm used by my image processing system. He directed me to places where I could find out how it works. It is an surprisingly simple matrix mathamatics process repeated over and over very fast by the computer. It brings to mind how mathematics is an elegant language of creation. Is it *the* language of creation ?

Here are my current thoughts on the apocalyptic images revealed from my cloud photos. Elsewhere I show that the nano-particle Chemtrail sprayers in the upper atmosphere are not the regular human aircraft many think. What if the nano size of the particles and the wavelength the aliens exist at played a role in revealing the aliens. Could the aliens be self disclosing because our governments are taking too long to reveal to the people the existence of the aliens and other important events about to occur?   www.cloudofIron.com

This next 'cloud' is familiar. Besides looking like the famous "Enterprise",
It also showed up in some photos I took of the Sun through a Solar Filter. Those pictures showed what appeared to be fireballs from the direction of the Sun. This object was in the lower right of the frame.

Stardate: 2013-May-23 Starship Enterprise shaped object. (StarTrek)   www.cloudofIron.com

I wasn't sure if I wanted to include this as I thought it would reduce my credibility.
Since it was an honest photo, I humbly submit it for your perusal....

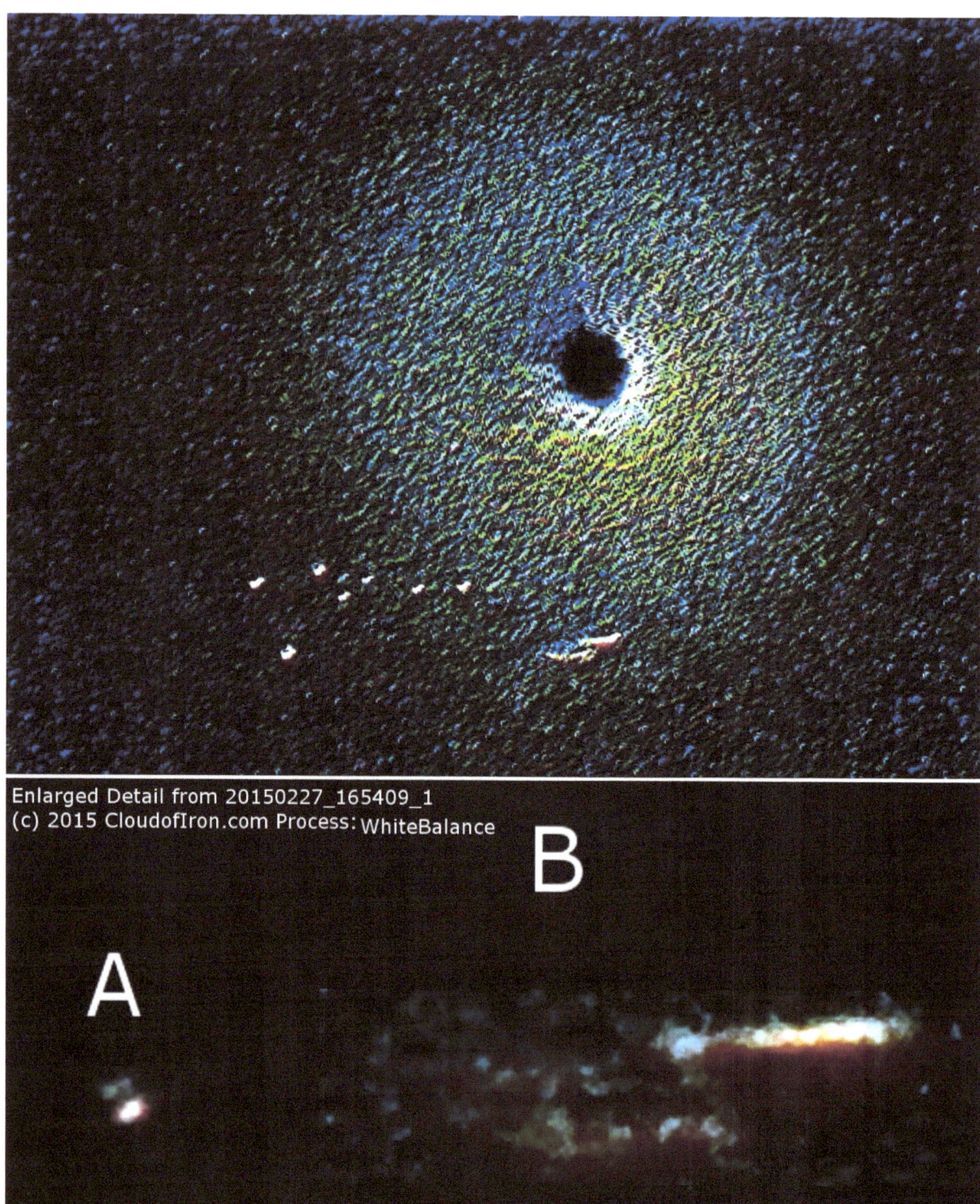

Enlarged Detail from 20150227_165409_1
(c) 2015 CloudofIron.com Process: WhiteBalance

2013-NOV-11 7:18am The Sun Rises over Queens, New York City. This photograph was taken through a double glass window. The version on the right, processed with edge detection software show a somewhat different picture. Above the Sun, is a prone figure with head at the 12 o'clock position. Below, and in back of him is a detailed head and possible shoulders with a hat or part of a helmet on. The number 19 is visible in both the original and the processed image. A bird stands like a blue rimmed hole in the sky. A figure stands up in this hole, smiling, facing a dark yacht with wings – upon which his chin rests. There are hints of other figures under the helmeted one. In addition to the green ball and red circle (Solar projection artifacts?), at building top and tree level are many items that I cannot see with my bare eyes. On top, parallel to the chemtrail is a red man, arms outstretched.

The 'Tauroctony' of the Mithraic Mysteries.

Little is known of the Cult of Mithras. The scene depicted here is found at the cult places of worship throughout the Roman World dating to around the first and second centuries. These scenes found in caves or places made to look like caves show Mithras slaying the Bull. Why does Mithras not ever look at what he is doing? Does he think something is coming? Is he telling us that the message is not just that he slays the bull but to look to that part of the heavens represented by the astrological symbols therein?

A photograph taken in June 2014 during a heavy thunderstorm in central Queens, NY

A man stands on a cloud. Who creates these cloud sculptures? Why?
Processing: Top of photo: Manual Col/Contr
Bottom of photo: White Bal.

"RESTARTING SATURN", a story.

In each volume of "Seen Flying the Skies of NYC,,," I am including a fictional story I have written, a 'what if', that lets my imagination run with some of the ideas and concepts I encountered, photographed or discussed in that volume. This is fiction, from the dark recesses of my mind and any resemblance to real life is merely coincidental. Please enjoy the story.

Man's "Golden Age" ruled by the god Saturn, here, on a cloud.

GOLTZIUS, HENDRICK (workshop)   (Muhlbrecht Haarlem 1558- 1617)
C. 1589. copper etching . 17.5 to 25 cm . Watermark (head jester with seven bells), Bartsch III, 33.

There was a time, "The Golden Age", the *first* age of Man, where the fruits of the Earth were enjoyed with little labor. In those days, men were just and lived without fear. It was the time when Creation was ruled by Kronos/Saturn and it is said there were two Suns. The one we call our Sun today, rose and set as it does. The other, a star, which was to become the ringed planet we now know as Saturn.

"RESTARTING SATURN"

Oxford Dictionary: **shamal**
Syllabification: sha·mal
Pronunciation: /SHəˈmäl/

Definition of *shamal*:
noun
A hot, dry northwesterly wind blowing across the Persian Gulf in summer, typically causing sandstorms.

Origin
Late 17th century: from Arabic *šamāl* 'north (wind)'.

Project Shamal: A secret, compartmentalized program of cooperative efforts between leading Earth American, British and Saudi scientists, their Governments and an ancient advanced civilization originally from Earth and now living on a number of planets in the direction of the Orion Constellation. The working relationship began in early 1948. The suggestion of the name 'Shamal', was that of the aliens; in their language it means 'to tie together', 'to bind', as in a partnership. Mankind and its great institutions of higher learning welcomed the name. As Earth Federation Director Alhazen reminded attendees at the April 2001 annual retreat, 'Shamal' - Arabic for the 'North Wind' - was an apt metaphor for the new paradigm, clearing the old to make way for the new.

The organizational chart was basically simple although there were lots of personnel. Each US Armed Service contributed a board member as did the Center for Disease Control, Central Intelligence, Homeland Security, Energy and NASA. The British and Saudis contributed exactly one member each leaving no doubt whatsoever as to who was in charge. The confidential 'Daily' report had key proposals, problems, solutions and negotiating points on the desk of the NSA Director each morning. A   much abbreviated summary version went to the US President each month. It's long been rumored that the NSA was providing certain civilian 'captains of industry' their own updates - some daily.

There had been much controversy as just how they should be treated, once the aliens had made themselves known. They could physically appear at will in a form just like us Earth bound humans, in fact that was one of their two natural states.   Over the billions of years after they had left Earth they had, with the help of others, the revered 'Ancient Ones', achieved a spiritual Nirvana that was still a far off goal for us here on Earth. They were spiritual beings that lived their immensely long lives by the strictest code of ethics and morals. Their second state of being was what we would call a metaphysical 'entity.' Everything in the human experience in the areas of psychic phenomenon was real and "they" were living proof of it.

Much of the controversy amongst men, had centered on the philosophical question of how to regard them. It seemed obvious to all that humanity had considered them as Gods over the millennium. Yet, they *had* started as a kind of human long ago and both 'Humanism' and arrogance were in vogue. Should they be worshiped? Would the aliens tolerate their creation trying to deal with them on equal footing? How should nations protect themselves? Should the public be told? That mankind should even have the audacity to consider these questions showed how far we had come from our humble roots. As in the past when confronted with a new potentially threatening situation, leaders turned to their generals for advice. Are you surprised that the road taken then was of a military nature? There was some shuffling and intrigues between countries anxious to have a 'special relationship' with the aliens. But, all did agree that it was best to maintain secrecy from the public. Who knows what the masses would do confronted with the first true information they had ever been given.

The aliens indicated that they preferred dealing with a single contact rather than one for each nation. A single contact? The United Nations was only three years old. It was unclear it would last another three. This might require a single world government to effectively handle this wouldn't it? The entire subject vanished into the black hole that is the American military industrial complex. A new level of secrecy was invented. Ways were found to increase compartmentalization; until it reached levels of inefficiency previously never attained. Motions to activate the U.N. in this liaison role were repeatedly stalled by the Americans. Another year and the Americans were confident in their new roles as spokesmen for a planet. For some reason, the aliens were giving, just *giving* away the most incredible technologies. Below is a list of some of the initial scientific advances. By 1955, all of them had been implemented to varying degrees.

A, Antigravity

B. Full Cloaking (invisibility)

C. A small water based engine was developed that was so simple, elegant and efficient, that every scientist privy to its design slapped themselves on the head saying "How did we miss THAT?"

D. The 'Killer G' problem was solved, enabling the new transportation technologies. Theoretical physicists had long said that the practical limit on flying craft would not be materials science but would be due to the frailty of the human occupants. Now, the new designs which looked like and were in fact, flying saucers and enormous triangular shaped craft could go from zero to nine tenths the speed of light in a millisecond. They could make those sharp ninety degree turns at 50,000 mph without harming a hair on the occupants.

E. Photo lithography. The technology to create electronic circuitry on silicon wafers (and transistors)

F. A working design for the Internet that included new cellular technologies.

G. Technologies that allowed the new craft to refuel at full speed from moisture in the atmosphere or while passing close to a body of water.

H. Total Weather control, including cloud sculpting capability delivered anywhere on the planet as one of the features designed into the new flyers.

I. The new flying technology scaled up very well indeed. Some craft had been produced that were more than 1 mile long. These could hold tens of thousands of people quite comfortably. The universe had just gotten smaller.

J. The ability to project holograms that looked as real as the actual object they were an image of. It could be done without a visible 'searchlight' like beam of light.

K. Laser systems of unprecedented strength harnessing again, the energy in good old water. They could use light at wavelengths invisible to the human eye and even could transmit power over inter-planetary distances,

L. Almost fifty new types of crops and animals were paraded before the scientists. Potatoes that were immune to blight. Corn that grew in 4 weeks. These were the answers to the world's food problems.

M. The diseases that had decimated Mankind for centuries were vanishing.

N. Tuberculosis was almost eradicated. Next would be Polio.

O. The entire science of astrophysics was rewritten as it became clear that stars were not giant nuclear reactors. They were more like giant electrical dynamos harnessing the power and balance of the forces acting at the borders of accretion disks, the event horizons of tiny "black holes". There was a singularity at the center of each star.

And, every few months, important previously unknown parts of Mankind's history were revealed to the working group at Project Shamal.

It was truly the rebirth of the legendary '*Golden Age of Man*' when Saturn the star had illuminated the antediluvian skies as the Second Sun.   And it was happening so fast! There were huge ships even now, working on Saturn's immense ring system. Astrophysicists had calculated that creating a huge concave lens by carefully laying down a layer of nano sized reflective particles could be used with a 'new technology' laser system to re-ignite the giant gas planet. We can do anything!

We're even restarting Saturn!

Early planning   included keeping the entire operation secret from the public. There were many good reasons and the aliens had agreed. Could such big secrets possibly be kept from the public?

Many people don't know that the 'Manhattan Project', the World War II effort to develop an atomic bomb had over 100,000 people involved and that remained totally secret from the public; (if you ignore the fact that the Russians infiltrated the project.) The secrecy from the *public* was in fact, effective.

By 1956, there was a secret base on the dark side of the Moon with a working population of 15,000 US Navy personnel. Bases on Mars, its moon Phobos, a small outpost on Jupiter's Europa and a larger base on Saturn's moon Iapetus where an ancient city had been discovered beneath the surface of the probably artificial moon.

Meanwhile Mankind's official 'on the record' reach for the stars continued for the sake of the public. Astronauts were dangerously rocketed into space atop giant liquid fuel machines that were more like bombs than flying craft. Some were lost. This incremental reach for the stars was all true and actually happening but the real action had already been going on for decades in total secrecy.

Instead of the technological advances freeing humanity, the opposite occurred. Many of the technologies just stayed in the area of the Military. A few were allowed to leak out and humanity greatly benefited in those areas. But you can't eat computer chips and for the most of the world energy costs climbed ever higher. A new feudal system was in operation. The entire planet was being pushed to benefit a tiny percentage of the population.

The advanced technology given to humans in the mid twentieth century was already tested and worked perfectly. For example, over 175,000 flying platforms were constructed ranging in size from a two person size up to the immense dimensions of a 747.aircraft. With room for hundreds of passengers, not one ever failed or faltered. Not one was ever seen 'uncloaked' by the public. They were of a fairly simple design, a flat surface with an upside down 'U' shaped composite bar at the back that ran across the width of the platform. The larger craft had an equal performance record. Every single 'sighting of a UFO' over the years had been an intentional planned event, part of a 'disinformation campaign' to cover the expected problems with the new technology that just never occurred.

The devices were all constructed in vast underground facilities in the American Southwest.

Early technology gifts from the aliens had included sophisticated tunneling equipment that created huge underground systems to put the factories in. A complex system of deep four lane wide tunnels crisscrossed America, effectively an upgrade to the new Interstate Highway system that was being built topside. A submarine entrance allowing the largest submarines to enter a similar tunnel system from an underwater canyon that came close to land in Monterey Bay, California. Similar entrances were being planned at Alvin Canyon off of New England and off of Belize.   Naval bases springing up in places like Kansas raised a few eyebrows but nothing more.

Portions of both the land and underwater tunnels appeared to already exist and these existing works were being leveraged by the aliens who had provided suggested routes along with the equipment. Artifacts were being found by the tunnel engineers that just did not make sense. High technology different from current technology but disturbingly similar; being used in the same time period as primitive technology. For example, engineers had found some indestructible metal objects, a folding cutlery set and cooking equipment and short saw - in essence camping equipment, at what appeared to be a campsite in a cave like recess along one of the tunnels. They could they not identify the metal nor explain why the massive bones from the giant, long extinct wild boar had been brought down with rock pointed spears. Some of the spear points were still in the animal's bones. Forensics proved without a doubt that   the saw that had found at the site been used to reduce the animal to manageable size after skinning it. There were lots of mysterious anomalies such as this.

 Unknown to all but a few, was the fact that there were more secret plans that only an 'inner circle' were privy to. Another secret plan in operation, an 'all human plan'. It was for nothing less than an audacious grasp for total power and control on the Earth. This was the goal of having a single world government often referred to by the fatally paranoid conspiracy theory nuts as the *"New World Order."* The plan had been being executed patiently over centuries. The Conspiracy Theory 'Nuts' had not been too off base on this one. But it was nothing new. It was just an extension of Mankind's base instinct for power. It wasn't even really about the Money. It was about the POWER.

Actually, the aliens of course had known about it, expected it. It was normal for a primitive Type I world-civilization like that of the Earth's. On a farm, would you expect the pigs to not try to eat as much as they can get? Or, would the humans make the transition, to the behaviors associated with a Type II civilization? *That* was the hoped for next step in their inevitable evolution towards joining the 'Beings of Light' that were the Aliens.

Time passed....

In the 1930's, military scientists in England had heard undecipherable beeping constantly on a very low radio frequency. Direction finding equipment had generally located where the signals were coming from but with the day's technology, that's all they had. The signal was not coming from anywhere on the Earth, yet it was near the Earth. The signal seemed to be coming from space, actually from different areas almost in space. It was as if the radio transmitter was on a plane flying a circumpolar orbit around the planet. What made it difficult to accept was the speed of the 'plane.' It seemed to be circling the globe every twenty minutes. What was it? What kept it from flying off into space as it whipped around the Earth at a mind boggling speed of over 50,000 miles per hour? The aliens claimed to not be its owner. Finally in 1958 its image was captured digitally by a 'Net' of aircraft using the new technologies. That morning, the aliens had reminded the humans that while it was not theirs it was also not ours and that we should not attempt to touch it or bring it down. Further pressing revealed that the owners were the 'Ancient Ones.' The fast moving satellite became known as the 'Black Knight' because of its unusual shape. There was one of these orbiting every known inhabited world. Newly terraformed worlds, found their own Black Knight in orbit within 1 hour of the first settlers. It had always been like that. There was no more discussion of it but obviously it was some kind of surveillance device - maybe even a 'Plan B' doomsday device to end a planet gone awry. The humans left it alone.

In spite of this 'good behavior' on the part of the earthlings it was becoming obvious to all that there was something wrong. Some expectation hadn't been met perhaps? It was then that the 'dreams' began.

To this day, the topic is surrounded by a thick veil of secrecy and considered a deeply personal vision by all who experienced it in late 1957 and early 1958. Every single head of state on the entire planet and members of their family had a strange dream. Every top officer in a major company and their staff and families. Leaders in religion, even some smaller key organizations and individuals destined for future greatness had the experience.

It began like this…Large winged beings would suddenly appear, interrupting whatever dream was being dreamed by the person. The person would become filled with a feeling of joy and peace, a feeling that no matter what happened, it would be OK. Perfect acceptance. Then, like a movie playing, they would see a tiny dot, it would grow larger. Now they could read a label next to it. "Hydrogen Atom Proton Nucleus." It grew, morphing into a larger dot labeled "Hydrogen Atom." Softly playing music that could only be described as religiously inspirational provided the background against which the object changed into another slightly larger, slightly more complex one.

The sleeping President, Premier or Tribal Chief watched as the parade of objects got increasingly larger. Atoms to Molecules; molecules flowed into enzymes; enzymes into nucleotides. Now four strands appeared, alive, twisting around each other forming the double helical shape of DNA. As these three dimensional visuals exploded vividly in the minds of the great powers on the Earth a gentle voice wove in and out of the music. "… and *that* is five hundred times larger…" Viruses, bacteria, now a tiny influenza germ, and then an insect. Coming at the viewer faster and faster, the parade of life only briefly showed Man.

Plants, Dinosaurs, Blue Whale, buildings, forests, mountains, continents and seas. As the size increased, the 'spell' upon the dreamer grew deeper still. As a frightened child is deeply comforted by being held in the Mother's loving embrace the sleeping men and women whose combined decisions affected the lives of billions were filled with a sense of peace; a primeval serenity, the certainty that all is well, all is going as it should.

The objects presented directly into their visual cortex continued to increase in size. The planets, the Sun, Stars twice the size of our Sun. Stars fifteen times the size of our Sun. A sense of incredible awe began to resonate for the dreaming heads of state. The music majestically rising in volume. Now the massive star: Rigel, Betelgeuse, VY Canis Majoris - "…over 2,000 times the size of your Sun…" Now the march included vast nebula; the "Cats Eye Nebula", the "Horsehead Nebula," the "Pillars of Creation", the "Milky Way"; Spiral galaxies twice the size of the Milky Way. Spiral Galaxies 1000 times the size of the Milky Way. Groups of Galaxies, Incomprehensible numbers of galaxies made up of uncountable numbers of stars. Each a Sun; each with planetary systems. Billions of planets. Life was everywhere!

Then, the parade of life abruptly stopped. A dark silence filled the dreamers. The dream had changed; but none awoke just yet, there was still much to learn…

The dream resumed as if from a momentary glitch. The dreamers were pervaded by an increasing sense that what was out there was not limited to what could be seen with their poor mortal senses. All around the dreamers now, twisted quantum force lines, wove a tapestry through inter-dimensional spaces. There was no such thing as empty space! Everything was teeming with life. Space itself gently expanding and contracting - alive and breathing. Right here and now, there were intra-dimensional beings interacting with humans in ways we couldn't even start to understand. Each dreamer was then brought to a sickening stop as they were struck with the realization of the uncountable numbers of souls living in these dimensions that had vanished forever in the split seconds around the detonation of each atomic bomb. Even the test explosions had caused billions of living beings to just vanish. They had been living in places we couldn't see. In impossible mathematical probabilities of inter-dimensional space. Many of the dreamers awoke with tears streaming from their eyes. Heads of State, renowned for their ruthlessness; some for their cruelty. All remembered the aliens at Project Shamal's meetings pleading for the nuclear experiments to stop. All, now filled with a deep guilt, remorse and sorrow. They opened their eyes ready to serve new masters in a new world order.

This dream, or vision, was the final calling card of the Annunaki. These humanoid aliens had known the earth from its earliest days. It was they who had seen Earth's potential as a jewel craftsman sees the diamond, gazing into a rough stone. It was they who waited for the newly formed planet to cool so they could terra-form it. It was they, along with one of the 'Ancient Ones' who finally gave the new planet its greatest blessing of all; Life.

The Annunaki had been with us in various forms since our beginning. Grown up with the Earth as their own and had nurtured her through all the things that can happen to a planet. Comets, asteroids, even defending it from other civilizations intent on taking the valuable resources of this beautiful ocean world. The Annunaki had a special relationship with this planet. Now, they also had some bad news…

It was Kiran Veeve who first mentioned it at the weekly Project Shamal working group in February of 1959. He just gave the human Science Officer a set of coordinates that defined a small square of space in the Orion Constellation. It was the same spot pointed to by the "John Gesture" in classical art. And the same point that Mithras's dagger entered the bull in every single "Tauroctonic" representation of the Mithraic Mysteries. It was the same spot that Stonehenge had been designed to watch and the same spot that the Pyramids of Egypt had been built to mirror. It was the place in the heavens described symbolically in Ezekiel's Vision and the Story of Perseus and the Gorgon. It was the exact same place described symbolically in almost every one of the world's mythologies.

A telescope in Chile was setup to watch the spot for the prescribed time as the alien had suggested. The news was bad. Very bad. There were dim objects there. So dim in fact, they would have been missed had the telescope not been trained on the spot for a week. And the scientists had seen it. What could be done? Was it too late?

What the scientists had seen was chilling. A brown dwarf star, about a third the size of the Sun. But that was not all, that star was the center of a solar system just as our Sun is the center of our Solar System. Enough data was gathered to get a complete orbital calculation. The entire system which included at least one huge planet four times the size of Earth and a debris field that trailed the system for almost 200 million miles was headed to keep its 3,600 year appointment with our Sun. It was already visible to patient astronomers, by 2011 it would be visible to any sharp eyed humans who knew where to look. The earthquakes and many other problems could start as early as 2005. The star and its companions would be on their way out of our Solar System by 2015 leaving Earth to deal with a debris field as wide as the Earth's orbit around the Sun.

This was our Sun's 'evil twin', a long hypothesized binary companion. Sometimes called 'Planet-X' when it was first searched for in the 1980's. This explained the orbital perturbations that had been observed in the outer planets Uranus and Neptune. It explained the course deviations Voyager missions had experienced as they left our Solar System.

It also explained the rise and fall of civilizations here on Earth.

A devastation so complete that legendary continents and places such as Atlantis could not be found. *Wiped from the face of the Earth.*

This was why the archaeological record was not finding any data to support evolution. The planet was being reseeded by aliens, if needed, after the disaster. Also over the years, archeologists had found fully developed civilizations appearing suddenly with full technology and then they vanished from the archaeological record without a trace. Every 3,600 years, 'Tabula rasa' as they say in Latin. "A Blank Slate." Unimaginable catastrophe on a planet wide scale. This was why in Hindu tradition, Lord Shiva destroyed all before Brahma could recreate the world. This was why the Mayans had predicted a new age. This was why the Native American Hopi's had a tradition that they had emerged from a hole in the ground into the creation of the Fourth World. *This* is what the Biblical book of Revelation was talking about.

This system of a star and planets is married to our Sun. All of the other planets including Earth are in roughly the same plane, like the top of a table; with the Sun at the center of the table. The flat plane of the table is the 'ecliptic.' The Sun's binary companion and its accompanying planets approach our Solar System not from the plane or same line of sight that we are on but from beneath the table and *behind* the Sun. As the Earth orbits the Sun, this system is also orbiting in a very elliptical orbit and it stays behind the Sun obscured from view by the Sun's brilliance until it has come so close that it breaks through the glare of the Sun. When it does break through the Bible says the sight will make men die of fright. Since we are all to die *anyway*, those who die having seen the 'Return of the Sun' or 'Second Coming' or the Winged Disk, or the "Destroyer" of the Egyptians should consider themselves blessed to have been able to see how it all ends. Don't you agree?

Man would survive; at least some would - enough to start again…The Sumerians and many other ancient civilizations had recorded this event when it happened previously.. We had read about it in their cuneiform writing on clay tablets. We had heard of it echoed in every single myth and legend ever passed down the generations. But, had we heard it too late this time.

 As the Annanuki had done countless times before (and fully recorded in the Vedic literature of India amongst other places), the 'rescue' technology had been given to mankind over fifty years before the arrival of the "Destroyer." Would man use this new information and technology to save every bit of DNA? Would he take all on the Ark? Would he *build* an Ark?

Fear makes drowning men climb up on their rescuer, drowning all. I am hungry. I will take all your food. Never does it occur to all to share. Is man without hope?

The 1960's through 2005 were very busy. Armed with this news of an impending catastrophe, decisions had been made. Tough decisions. The caverns used to build the flying craft were re-purposed as refuges. They were stocked with food, clothing. Weapons. Communication systems such as cell systems, and normally private information such as insurance records, were co-opted to be used for surveillance and control. Devices to take over remote control of motor vehicles by police Agencies were placed in all vehicles air, land and sea. Taking photos on film became a thing of the past replaced by digital photography. You couldn't even find a place to develop film by 2014. They would keep the negatives if you did. You would store your photos on your computer where they could be scooped up nightly and filed inside mountains; no not to keep records but to identify bodies and enemies of the state. The cell phones we all carried became beacons that can advertised where we were, their cameras being used to get close up tactical views of the battle. In fact, because of the monumental impact on the collapsed medical system, insurance policies were changed by decree in the last years to protect the related investment companies that like everything else in the financial system were ultimately part of the 'Machine.'. Huge databases on the population were merged and moved to areas where they were placed deep underground to help guarantee that power remained with those that currently had it. Instead of trying to get the masses to team up and prepare for rebuilding, plans were made to introduce martial law with contrived provocation events and military level weapons were issued to local police to handle the resultant rioting. Incredibly sophisticated mind control systems were implemented. All of them designed to get the populations attention after the Dark Star and Nibiru become visible. A population barely controllable even with the '3 and 3' legal system to handle serious uprisings. '3 and 3'? Yes, three Judges with final authority to find a party guilty with a single punishment;

Death – no appeals. Sentences to be carried out within three days of arrest. Wholesale execution of prison populations and 'vanishing' of troublemakers. If at that moment of hopelessness, an image of the Messiah appears in the sky a mile high and speaks to the people they will go (went the thinking). "Pack one bag per person and board the trains peacefully, you will be returning - the Army will protect people's property. There is hot food at the centers, and hot showers too. Quick now, lest you be trapped on the coast after others have been moved to a safer location. "

A simulated meteor strike in central New York City, between the Russian Jewish neighborhood of Rego Park and Hispanic Jackson Heights was even planned to move things along. During an overcast work day, it could be done with explosives supported by high tech holography and the releasing of some meteor footage from a satellite or plane above the cloud cover. *It had been done before...*

The Black Knight Satellite remained without additional comment, orbiting as it had since the day it was put into operation.

The Annunaki were beside themselves with despair. It was their home system that trashed the Earth every 3,600 years. Was it the fundamental nature of mankind that resulted in an appreciable self-rescue of mankind only about once every ten cycles no matter what they did to help. There was not much they could do, nor much they should do - mankind had made his bed, now he would sleep in it. They had seen it all before. Those who thought they could survive in underground tunnel systems usually found that the tunnels became their tombs. The deeper they were the more likely they would live through the initial contractions but they would be unable to tunnel out either because they were too deep or the underground system filled with oil or lava. The survivors were usually the most primitive. The Australian Aborigines would probably be ok. The Native Americans such as the Hopi would also probably make it; emerging into their 'Fifth World.'

The Annunaki, compassionate with their creation yet again, decided to do the 'firmament thing.' It had worked before and would work again. They set their flying machines to look like large jet aircraft and loaded up the canisters with the Nano particles of metallic aluminum and barium and other chemicals. Designed to be able to cross the blood-brain barrier, they dull the brain, relieving anxiety. Designed to reduce lawlessness the process also creates a spherical layer around the Earth. This layer would obscure the approach of the Dark Star and Planet Nibiru until right before whatever was going to happen happened. Night and day their flying craft traveled the stratosphere weaving the firmament layer in the last couple of years before the event. Some of the more aware public called the wispy trails high up in the sky 'Chemtrails.'

The Government knew what was happening and even created another misinformation campaign to lead the public to think it was a global-warming initiative that had been kept secret because it contained chemicals that could cause cancer - but since it was for the good of Mankind they were doing it anyway. The fact was, it was being done by alien ships for alien purposes; one of them being compassion for their creation. Something certain humans would not know anything about.

The Black Knight Satellite must have overheard something. It started sending out a new repeated pattern over the same radio channel it had been transmitting on for millions of years. This was recognized immediately by an analyst at the National Security Agency as simple Morse Code sent by someone who didn't think an extra letter should be wasted on these humans; And the telegram style message read:

HERD U MAY HV PRBLM -(STOP)- R U STIL PLANNG TO RESTRT SATRN? STOP

Speaking of Saturn, Additional photos of Saturn's Hexagonal North Pole coincided with an accidental 'selfie' at 01/18/2014 1:18am below

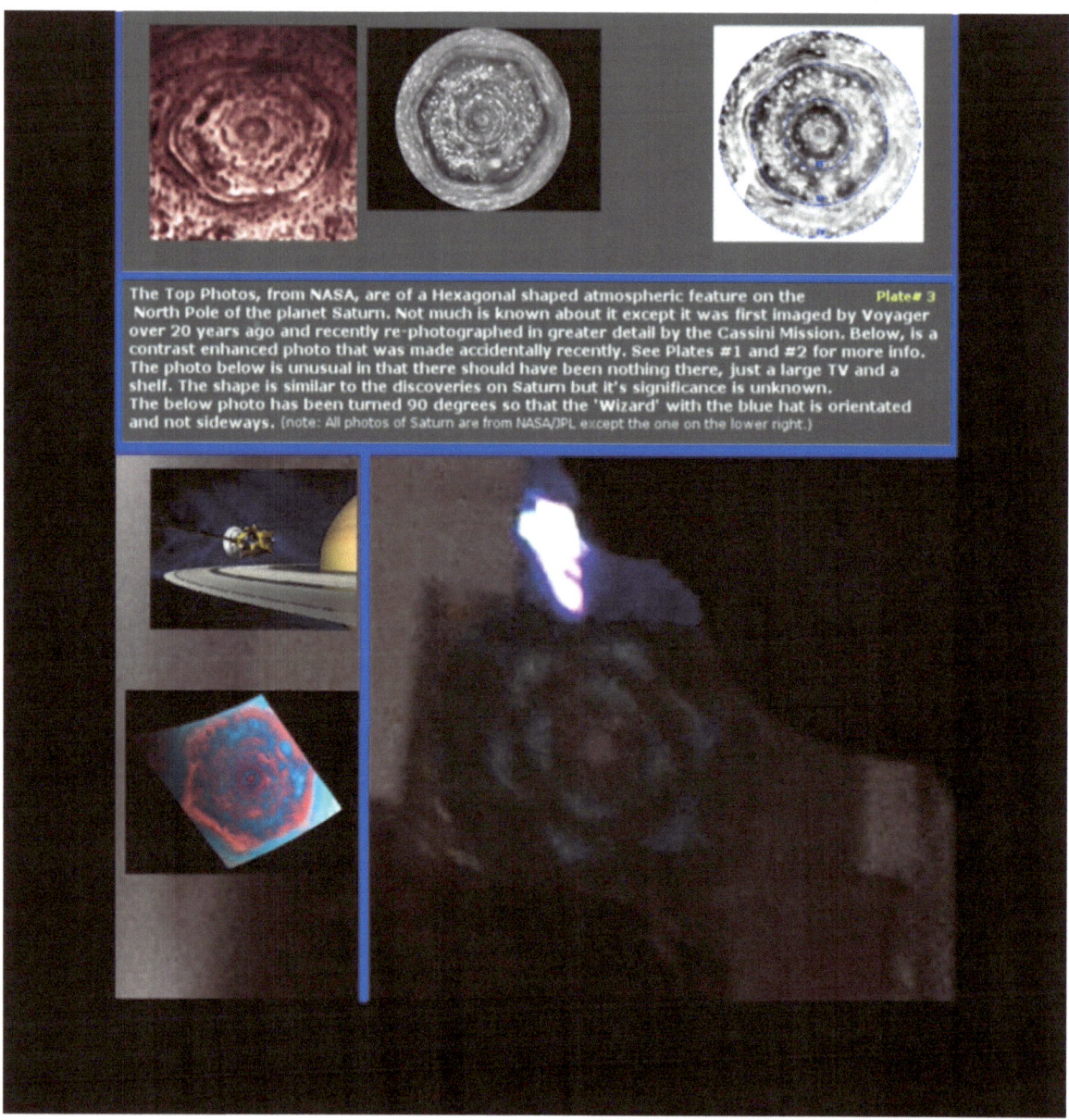

The Top Photos, from NASA, are of a Hexagonal shaped atmospheric feature on the **Plate# 3** North Pole of the planet Saturn. Not much is known about it except it was first imaged by Voyager over 20 years ago and recently re-photographed in greater detail by the Cassini Mission. Below, is a contrast enhanced photo that was made accidentally recently. See Plates #1 and #2 for more info. The photo below is unusual in that there should have been nothing there, just a large TV and a shelf. The shape is similar to the discoveries on Saturn but it's significance is unknown. The below photo has been turned 90 degrees so that the 'Wizard' with the blue hat is orientated and not sideways. (note: All photos of Saturn are from NASA/JPL except the one on the lower right.)

2015-Mar-17 "The Destroyer's Advance Party Arrives"   www.cloudofIron.com

2015-Mar-17 NYC "... and I hid. Then, I heard a bump on the roof. It had stopped; it was looking for something..." www.cloudofIron.com

2015-Mar-17 "Look! Up in the Sky!, It's a bird...No, it's a plane, No, It's the Annunaki! from the planet Nibiru !" Note Book Reading figure in right sky.   www.cloudofIron.com

The survivors of this event will be looked on as the first civilizations; much as we look back at the early Mesopotamians. There will be a book, similar to our Bible to pass down knowledge without sentencing future generations to the hopelessness of rebuilding civilization only to have it all wiped out as the cycle continues. As our current Bible does, it will speak of a 'Firmament' that stretched from the Earth to the heavens. The philosophers then, will say that light from the Sun, Moon and Stars emanated through holes in the Firmament.   Then one day, our unseen godlike friends will crack the shell of the egg, and Astronomy will be reborn. Other restrictions to our descendant's imaginations will be lifted. Now though, a new world order is ushered in by these apocalyptic images and the photographs of the 'aircraft' that are noiselessly crossing our skies. Hard at work, busy building, a fresh Firmament for a New Age; the one long awaited by many cultures – The Fifth World of the Hopi Indians.

-   Argus Witness   witness@cloudofIron.com
    May 10, 2015 New York City

Science Fiction Predicts Real Science

There is a long tradition of Science Fiction becoming reality some time later. There used to be a genre 'fantasy' , but then writers discovered that along with the advances of the Industrial Revolution came a new technique in writing. It looked like Man could do anything. Some things just took a little longer. Using the plot device of inventing a gadget or technology so the story works, writers like Jules Verne became designers of the world of tomorrow. Their stories represent the dreams of a world where anything can be done.

There was a fictional story in" Seen Flying the Skies of NYC and Mars", vol1 about the surprising discovery of a human face on a comet.

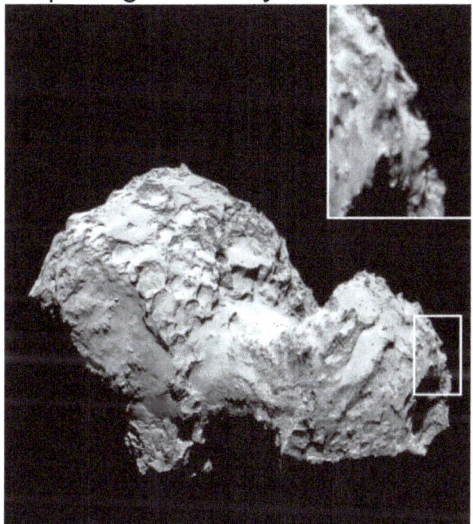

Months later the Rosetta Mission released a photo of possible face on the comet 67/P.

This is different from the story line in the book "Seen Flying…" but it is an interesting coincidence. As an aside, the three letter abbreviation of the organization that released the photo within the European Space Agency Just happens to be my three initials. I thought that was odd.

Man has come a long way, a *very* long way in science since the likes of Jules Verne.
I thought I overheard someone talking about
Restarting Saturn….

www.ingramcontent.com/pod-product-compliance
Lightning Source LLC
Chambersburg PA
CBHW050757180526
45159CB00003B/1493